Deforestation *and* Reforestation in Namibia

Deforestation *and* Reforestation in Namibia

The global consequences of local contradictions

Emmanuel Kreike

Markus Wiener Publishers
Princeton

Cover photo: The elaborate palisades that mark north-central Namibian homesteads contribute to deforestation, but, at the same time, farms have been prime sites for reforestation with fruit trees. A baobab tree dominates in the center of the photo next to the palisade of senior headman Mathias Walaulu's Okalongo homestead. To the left rises a real fan palm, and in the background, marula and birdplum trees tower over the millet fields. (Photo by the author, 1993)

For information, write to Markus Wiener Publishers
231 Nassau Street, Princeton, NJ 08542
www.markuswiener.com

Library of Congress Cataloging-in-Publication Data

Kreike, Emmanuel, 1959-
Deforestation and reforestation in Namibia : the global consequences of local contradictions / Emmanuel Kreike.
p. cm.
Includes bibliographical references and index.
ISBN 978-1-55876-497-2 (hardcover : alk. paper)
ISBN 978-1-55876-498-9 (pbk. : alk. paper)
1. Deforestation—Namibia. 2. Reforestation—Namibia.
3. Namibia—Environmental conditions. 4. Environmental policy—Namibia.
5. Global environmental change. I. Title.
SD418.3.N34K74 2010
333.75096881—dc22

2009033989

Markus Wiener Publishers books are printed in the United States of America on acid-free paper and meet the guidelines for permanence and durability of the Committee on Production Guidelines for Book Longevity of the Council on Library Resources.

Contents

Acknowledgments

This book is the outcome of a long journey that began in 1990 at the Department of Forestry at Wageningen University under the guidance of Adriaan van Maaren, Marius Wessel, and Freerk Wiersum. It also owes much to my mentors at Yale University, especially Robert Harms, James Scott, and the late Robin Winks. I was fortunate to receive valuable feedback at a variety of wonderful venues where I presented drafts of the work, including meetings of the African Studies Association; the Agrarian Studies Program at Yale University; the Davies Center of the History Department and the Princeton Environmental Institute, both at Princeton University; the Dutch CERES Research School for Resource Studies for Development; and the European Environmental History Association.

The research in Namibia that informs the book was an intense and rewarding collaborative project. It was made possible by a doctoral dissertation fellowship from the Social Science Research Council, New York. Thanks to the linguistic, social, and geo-spatial skills of Jackson Hamatwi, a former teacher at St. Mary's High School at Odibo, I was able to meet with many highly knowledgeable inhabitants in north-central Namibia who shared their experiences with and insights about environmental change with me. They include Julius Abraham, Kaulipondwa Tuyenikalao Augustaf, Abisai Dula, Philippus Haidima, Kulaumoni Haifeke, Helaliah Hailonda, Nahango Hailonga, Mwulifundja Linekela Haiyaka, Alpheus Hamundja, Hendrik Hamunime, Helemiah Hamutenya, Kanime Hamyela, Shangeshapwako Rachela Hauladi, Juliah Hauwuulu, Israel Hendjala, Francisca Herman, Monika Hidengwa, Petrus Shanika Hipetwa, Matias Kafita, Moses

Kakoto, Malita Kalomo, Joseph Kambangula, Gabriel Kautwima, Mathias Malaula, Magdalena Malonde, Islael Mbuba, Petrus Mbubi, Kaulikalelwa Oshitina Muhonghwo, Moses Mundjele, Helivi Mungandjela, Joshua Mutilifa, Kalolina Naholo, Helena Nailonga, Timotheus Nakale, Ester Nande, Paulus Nandenga, Matteus Nangobe, Emilia Nusiku Nangolo, Elisabeth Ndemutela, Werner Nghionanye, Joseph Nghudika, Louisa Palanga, Pauline, Lea Paulus, Marcus Paulus, Vittoria Petrus, Lydia Polopolo, Twemuna Shifidi, Erastus Shilongo, Johannes Shipunda, Joseph Shuya, Selma Tobias, Salome Tushimbeni, Paulus Wanakashimba, and Maria Weyulu.

Many in Namibia in addition to the elders who were interviewed welcomed me into their homes and shared their ideas with me. I would especially like to thank Dr. Peter and Jane Katjavivi, Bishop and Sally Kauluma, the late Michael Hishikushitja, and Joseph Hailwa, the Director of the Department of Forestry.

At Ogongo Agricultural College in Namibia, Haveeshe Nekongo, Arne Larsen, Carlos Salinas, and their colleagues and students contributed greatly to the project, not in the least through assisting in developing and administering the Ovamboland Multi-Purpose Investigation for Tree-use Improvement (OMITI) household survey. The support of the Namibian Directorate of Forestry, the Dutch Embassy in Namibia, and IBIS-Denmark made the OMITI survey financially and logistically possible.

Various archives, including the Historical Archive of Angola in Luanda, the Archive for Overseas History in Lisbon, the Archive of the Missionary Holy Ghost Congregation in Paris, the United Evangelical Mission in Wuppertal-Barmen, and especially the National Archives of Namibia in Windhoek provided rich data on the environmental past of the Ovambo floodplain and the surrounding region. The late Brigitte Lau and Werner Hillebrecht, respectively the former and the current director of the National Archives of Namibia, were immensely helpful.

I would also like to thank my colleagues at Princeton University, including Robert Tignor, Peter Brown, William Jordan, Angela Creager, as well as those beyond the Orange tower, including William Beinart, Peter Boomgaard, Steve Feierman, Peter Geschiere, K.E. Giller, P. Hebinck, Susanna Hecht, Andrew Isenberg, Robert Papstein, Petra van Dam, J.W.M. van Dijk, and Louis Warren, whose moral and intellectual support has been very important. I am also grateful to Dick Foeken and Jan-Bart Gewald of the African Studies Centre (Afrika-

Studiecentrum), Leiden; to two anonymous readers commissioned by the ASC and Brill; as well as to Gavin Lewis, Carol L. Martin, Joed Elich from Brill, and Markus Wiener from Markus Wiener Publishers for their invaluable advice in turning the manuscript into a book. Tsering Wangyal Shawa, GIS Librarian at the Geosciences Library, Princeton University, helped me with the maps.

Sections of earlier versions of chapters 1, 7, and 8 appeared as "The Nature-Culture Trap: A Critique of Late 20th Century Global Paradigms of Environmental Change in Africa and Beyond," in *Global Environment: A Journal of History and Natural Sciences* 1 (2008): 114–145. A part of chapter 6 appeared as "De-Globalisation and Deforestation in Colonial Africa: Closed Markets, the Cattle Complex, and Environmental Change in North-Central Namibia, 1890–1990," in *Journal of Southern African Studies* 35(1) (2009): 81–98. An earlier version of chapter 7 appeared as "The Palenque Paradox: Bush Cities, Bushmen, and the Bush," in A.C. Isenberg (ed.), *The Nature of Cities: Culture, Landscape, and Urban Space*, Shelby Cullom Davis Center for Historical Studies, Princeton University, Studies in Comparative History (Rochester, N.Y.: University of Rochester Press, 2006), pp. 159–174.

Light is critical to creating and sustaining physical and intellectual life. I am grateful to my parents Hermanus Kreike and Grace Kreike-Tak, my teachers, and my fellow students for instilling me with a love for study and to my grandparents Paulus and Adriana Tak for instilling in me a love for the land. My spouse, Dr. Carol Lynn Martin, and our children Hermanus Clay and Eleanora Grace, are my suns and my moons. It is to Carol that I dedicate this study.

Studies [illegible] to two anonymous readers commissioned by the ASC [illegible] Robin [illegible] Lewis [illegible] Martin, [illegible] Ellen [illegible] Bell [illegible] Weber [illegible] Publisher, for their [illegible] advice [illegible] Walker [illegible] GIS [illegible] at the Geos[illegible] Library [illegible] helped me with maps.

[illegible] of chapters [illegible] and 8 appeared as "The Nature–Culture [illegible] Paradoxes [illegible] Change [illegible] and Beyond," [illegible] Global Environment: Journal of History and Natural [illegible] Sciences [illegible]. Part of chapter 6 appeared as "[illegible] Globalisation and [illegible] Closed Markets: the Cattle Complex and Environmental [illegible] in North-Central Namibia, [illegible]," in Journal of Namibian Studies [illegible] (2009) [illegible]. An earlier version of chapter 7 appeared as "[illegible] Paradox [illegible] Cities, [illegible] in the Bush," in [illegible] Cities [illegible] Studies in Comparative [illegible] (Rochester, NY: University of Rochester Press, [illegible]).

[illegible]

Abbreviations

A233	Accessions: Chapman Collection
A450	Accessions: Carl Hugo Linsingen Hahn Papers
AGCSSp	Archives Générales de la Congrégation du Saint-Esprit (Central Archives of the Holy Spirit Congregation, Paris)
AGR	South West Africa Administration: Directorate of Agriculture
AHU	Arquivo Histórico Ultramarino (Archive for Overseas History, Lisbon)
ANC	Assistant Native Affairs Commissioner
ANCO	Assistant Native Affairs Commissioner Ovamboland
AVEM	Archiv der Vereinigten Evangelischen Mission (Archives of the United Evangelical Mission, Wuppertal-Barmen)
BAC	Bantu Affairs Commissioner
BOS	Bantu Affairs Commissioner Oshikango
CEM	Church of England Mission (Anglican)
CNC	Chief Native Commissioner
CNDIH	Centro Nacional de Documentação e Investigação Histórica (National Center for Historical Documentation and Research, Luanda)
CU	Cattle Unit
FMS	Finnish Mission Society
GRN	Government Representative Namakunde (Neutral Zone)
GVO	Government Veterinary Officer
NAN	National Archives of Namibia (Windhoek)
NAO	Native Affairs Ovamboland
NC	Native Commissioner
NCO	Native Commissioner Ovamboland
O/C	Officer Commanding
OMITI	Ovambo Multi-Purpose Investigation for Tree-Use Improvement
OTC	Ovamboland Trading Corporation
OVA	Ovamboland Administration
OVE	Ovamboland Economic Affairs Department

OVJ	Ovamboland Justice Department
RCO	Resident Commissioner Ovamboland
RMG	Rheinische Missionsgesellschaft (Rhenish Mission Society, Wuppertal-Barmen)
SAP	South African Police
Sec.	Secretary (for South West Africa, etc.)
SWA	South West Africa (Namibia)
SWAA	South West Africa Administrator
SWANLA	South West African Native Labour Association
SWAPO	South West Africa People's Organisation
UA	Union (of South Africa) Administration
UGR	Union Government Representative (Neutral Zone)
UNG	Union Government Representative Namakunde (Neutral Zone)
WAT	South West Africa Administration: Directorate of Water Affairs

Maps

Photos

1

Approaches to environmental change

The causes and consequences of environmental change have been hotly debated by academics, policy makers and the public at large since at least the 1960s. The prevailing literature focuses on evaluating environmental change against a baseline (such as pristine Nature) to assess whether the outcome is environmentally neutral, or one of environmental degradation or improvement. The most commonly used models analyze environmental change by highlighting one or more causative agents, including the so-called 'population bomb'; factors ascribed to colonialism and imperialist power struggles such as conservation policies and political ecology; ecological exchanges (such as those involved in the spread of diseases and in biological imperialism); economic globalization, for example the rise of capitalist markets; and new developments in technology (such as the use of firearms and steel). The case of north-central Namibia serves to demonstrate how these global models give rise to different and often contradictory interpretations even within a single approach that cannot be simply explained away as alternative readings or mis-readings of the same process. Twentieth-century north-central Namibia experienced dramatic deforestation *and* reforestation as a result of population pressure, and the area witnessed the *de*globalization of a precolonial global resource (cattle). Diamond's trinity of 'guns, germs and steel'[1]—the

[1] Diamond, *Guns, Germs, and Steel.*

unholy alliance of imperialism, ecological exchanges and technology—is certainly revealing. But global flows of microbes, firearms and iron technology shaped local southern African environments in nonlinear and unexpected ways. Some invasive germs caused deadly virgin soil epidemics in Africa, echoing the impact of smallpox in the Americas and paving the way for colonial conquest. But some of the invasive germs and guns and steel turned against colonialism, and caused colonial projects to veer sharply off course with unexpected environmental consequences. Whether caused by colonialism, population pressure, technology or invasive species, environmental change consequently should be understood to be multidirectional, involving multiple sub-processes with plural outcomes.

Despite path-breaking research in the past two to three decades, the study of local and global environmental change is constrained by the conceptualization of change as a singular process that is *both linear and homogenous.* Such a conceptualization creates two paradoxes that cannot satisfactorily be explained within the current frameworks and that are here referred to as the Palenque Paradox and the Ovambo Paradox.

Depicting environmental change in linear fashion within a Nature-Culture dichotomy has been rejected in theory. In practice, however, environmental change overwhelmingly continues to be assessed in terms of singular and exclusive degradation, improvement or stability / equilibrium outcomes. The degradation-or-improvement-or-equilibrium framework is derived from the modernization, the declinist and the inclinist paradigms, all of which share the premise that environmental change occurs along a single and irreversible Nature-to-Culture pathway.

The modernization paradigm posits environmental change as a progression from a primitive state of Nature to an advanced state of Culture, resulting in a state-controlled and scientifically exploited environment.[2] The declinist paradigm

[2] Goudsblom, for example, describes the progress of humankind through the domestication of fire in *Fire and Civilisation*. Nash argues that scientific conservation in the United States arose with the closing of the frontier, in *American Environmentalism*, pp. 69–112. But Grove traces the roots of Western conservation much farther back, in *Green Imperialism*. For critical overviews of the modernization paradigm see Richards, *Indigenous Agricultural Revolution: Ecology and Food Production in West Africa*, pp. 31–40; Blaikie and Brookfield, *Land Degradation and Society*, pp. xviii–xix. On modernization and Nature-to-Culture change, see Merchant, *Reinventing Eden*,

regards human interference in pristine Nature as a disturbance that leads to a downward-spiraling process of environmental degradation that ultimately might cause the destruction of ecosystem Earth.[3] In contrast to the largely pessimistic outlook of the declinists, and similarly to the modernizers, the inclinists are optimistic about humans' ability to mitigate the degrading effects of environmental change.[4]

But the presence of the urban ruins of Palenque and other 'lost cities' in the jungles of the Americas, Asia and Africa creates a paradox: if Culture once dominated the last remaining wildernesses of Planet Earth, how can these areas constitute (pristine) Nature? Moreover, environmental change is understood as a homogenous and singular process. The preoccupation with outcomes (and baselines) leads to a relative lack of appreciation for the dynamics of the *process* of environmental change itself. The liberal and often uncritical use of models from the natural sciences as an analytical shortcut to connect a particular environmental outcome to a specific past environmental baseline seems to make understanding the details of the process of environmental change less urgent. After all the models appear to explain how the outcome resulted from the baseline. Disagreements about the trajectories and the outcomes of environmental processes were attributed to different interpretations or misreadings of what was essentially the same process. Ambiguities in the process and the outcome, however, may also reflect contradictory subprocesses. Moreover, different sub-processes of environmental change may not be fully synchronized, suggesting the need to reconceptualize environmental change as a pluralistic set of processes. Descriptions of the late 1800s precolonial landscape of the Ovambo floodplain in the Angolan-Namibian border region closely match the area's late 1990s postcolonial appearance, suggesting little

---

pp. 20–186, and *Ecological Revolutions*; Thomas, *Man and the Natural World*; Bassett and Crummey, *African Savannas*, pp. 13–15; Worster, *Dust Bowl*, pp. 182–229; Stilgoe, *Common Landscape of America*; White, *The Organic Machine*, pp. 59–88; and Arnold, *The Problem of Nature*, pp. 1–74.

[3] Seminal works were Carson's *Silent Spring* and Worster's *Dust Bowl*. For global perspectives, see, for example, Westoby, *Introduction to World Forestry*; Chew, *World Ecological Degradation*; Williams, *Deforesting the Earth*; Myers, *Deforestation Rates in Tropical Forests*; Jepma, *Tropical Deforestation*. For a history of the declinist paradigm, see Merchant, *Reinventing Eden*, pp. 187–203.

[4] Bassett and Crummey, *African Savannas*, pp. 1–4. Henkemann, Persoon and Wiersum identify an emerging paradigm that stresses the human capacity for innovation in "Landscape Transformations of Pioneer Shifting Cultivators at the Forest Fringe", p. 55. See also Fairhead and Leach, *Reframing Deforestation*, p. 191.

change between the precolonial baseline and the postcolonial outcome. Yet, paradoxically, dramatic deforestation *and* reforestation marked the area's twentieth-century history.

Models of environmental change

Population pressure has been identified as a major if not *the* major driver of environmental change in the twentieth century. The process of population pressure-induced environmental change typically is depicted in mechanistic-linear terms. This is especially true in the case of (neo-)Malthusian models: population growth outpaces the growth of food production, leading to the overuse of natural resources and, eventually, environmental and societal collapse.[5] In other words, population pressure leads to the destruction first of Nature, and subsequently of the Culture that depends on it.[6] The environmental impact of population pressure, however, is contested: Boserup and others argue that population pressure triggers technological innovation and more effective natural resource management, making it possible to sustain larger populations without destroying Nature.[7] Both the Malthusian and the Boserupian models identify population pressure as a critical driver of environmental change, but they evaluate the outcome of the resulting process of change in diametrically opposite ways: Malthusian Armageddon versus Boserupian Utopia. It is undeniable that ecosystem Earth cannot support unlimited population growth, but the co-existence of two opposing views suggests at the very least that the trajectory of population pressure-induced change historically and theoretically is not predetermined or linear. As a result, the trajectory of population pressure-induced environmental change and its outcome may not be unambiguously negative (as in a declinist model) or positive (as in an inclinist model).

Moreover, until the 1940s or 1950s, in north-central Namibia, as elsewhere in Africa, population *movement* associated with a climate of political insecurity was a

---

[5] Malthus, *An Essay on the Principle of Population*; Ehrlich, *The Population Bomb*; Ehrlich and Ehrlich, *The Population Explosion*; Cleaver and Schreiber, *Reversing the Spiral.*

[6] A good example is Diamond, *Collapse.*

[7] Boserup, *The Conditions of Agricultural Growth*; Pingali, Bigot and Binswanger, *Agricultural Mechanization and the Evolution of Farming Systems in Sub-Saharan Africa*; Tiffen, Mortimore and Gichuki, *More People, Less Erosion.*

more critical variable than (natural) population *increase*. The impact of population and population pressure needs to be stressed not only as an abstract quantitative factor, but also as qualitative a factor that affects the environment through social and political processes.

A political ecology approach highlights the extent to which ideas, policies and practices related to the exploration and conquest of Africa and the administration of colonial empires are factors that shape the perception and direction of environmental change, a set of issues that Grove labeled 'green imperialism.' Colonial conservation and development priorities and projects shaped the non-Western environment physically and conceptually—often in very dramatic ways. The hunting and gathering of forest products, for example, was redefined as poaching when colonial administrators created game and forest reserves.[8] Moreover, the insecurity brought about by colonial conquest and the draconian punishment meted out to maintain colonial law and order—including conservation regulations—often caused massive population displacement with dramatic environmental consequences. In north-central Namibia, colonial officials increasingly enforced international and internal colonial borders to limit the movement of people and animals, and colonial policies restricted hunting and tree harvesting. Policies that had an even greater impact than proclaiming game reserves in the area, however, included disarming the local population in the 1920s and 1930s, and fencing the international, internal and game reserve boundaries in the 1950s and 1960s.

Old World biological invaders that accompanied European conquerors, most dramatically smallpox germs, sheep, cattle, horses and a host of plants, decimated precontact American indigenous human, animal and plant populations, destroying the local environment, and creating a Neo-Europe. Biological invaders coincidentally (and sometimes intentionally on the part of the conquerors) facilitated European conquest.[9] Overall, the biological invasion is portrayed as a unilinear, mechanical and progressive process of environmental change from Nature to Culture. The overall effect was the transformation of the earth into a unified

---

8 See, for example, Grove, *Green Imperialism*; Anderson and Grove, *Conservation in Africa*; MacKenzie, *Imperialism and the Natural World* and *The Empire of Nature*. See also Guha, *The Unquiet Woods*, and Peluso, *Rich Forests, Poor People*.

9 Crosby, *The Columbian Exchange* and *Ecological Imperialism*. See also McNeill, *Plagues and Peoples*.

ecosystem dominated by Western culture and Western (domesticated and/or 'weedy') species. In north-central Namibia, biological invaders included lungsickness, rinderpest and foot and mouth germs, as well as donkeys and horses. Lungsickness, rinderpest, foot and mouth and donkeys are ranked as major environmental scourges across southern Africa. The histories of biological invaders in the region, however, complicate linear narratives of environmental change. The impact of lungsickness and rinderpest in Africa mirrors the destructive impact of smallpox in the Americas. These germs caused dramatic domestic and wild animal losses, weakened preconquest societies and the environments they depended upon, and paved the way for colonial conquest. Lungsickness triggered the collapse of South Africa's Xhosa society and rinderpest had an enormously destructive impact across southern Africa in 1896 and 1897. In contrast, while reported outbreaks of foot and mouth in the 1950s and 1960s terrified colonial officials, the disease did not kill a single animal in north-central Namibia. Unable to eradicate foot and mouth or lungsickness, colonial administrations cordoned off the infected domestic and wild animal herds. The history of donkeys also undercuts linear and progressive Nature-to-Culture narratives. Unlike, for example, European sheep, which became a plague in Mexico, donkeys initially did not thrive in north-central Namibia at all, and their introduction does not adhere to the invasion-followed-by-explosive-growth-and-subsequent-implosion model that typifies biological invasion narratives.[10]

Economy-driven (market) models of environmental change are premised on the dichotomy of a precontact, local, barter-based moral economy that is replaced by a global market economy. The level of analysis is often abstract, with the driving force being identified as 'capitalism' or 'the market'. In Africa, it is often argued that population growth in semi-arid regions is accompanied by explosive increases in livestock numbers, resulting in overgrazing. These arguments build on the premise that human agency in livestock management is circumscribed by local culture or tradition, which in turn is determined by the limitations imposed by the natural environment. For example, in the 'cattle complex' model, cattle numbers increase beyond sustainable levels because cattle are not consumed

[10] Melville, *A Plague of Sheep*. Donkeys are native to Africa but not to the southwest of the continent.

or sold, but rather are hoarded as a symbol of status and wealth. Yet, hard evidence for the existence of either a cattle complex or a subsequent livestock population bomb resulting in overgrazing is as lacking in north-central Namibia as it is elsewhere.[11] North-central Namibia's cattle owners readily exported cattle across southern Africa and the Atlantic world before colonial rule, a practice that contradicts the 'precolonial' or 'traditional' origins of the presumed cattle complex phenomenon and the existence of a precolonial moral economy. This trajectory casts fundamental doubt on colonialism's reputation as an economic globalizing force. In north-central Namibia, colonial rule in fact in many ways deglobalized local economies. A linear mechanical model of market-driven overexploitation of natural resources thus seems too monodimensional.

A final important model of environmental change highlights the agency of Western technology. The model is premised on the assumption that new technology automatically creates its own demand because it is inherently and transparently superior, leading to the wholesale replacement of pre-existing technology. In the model, Western scientific technology (as a globalizing force) typically replaces local, traditional and primitive (labor-intensive) technology. Diamond argued that the West colonized the non-West (including Africa) rather than the other way around because of the West's early acquisition of superior technology, which paved the way for Western world dominance.[12] But sub-Saharan Africans produced steel before it was produced in the West and Africans resisted European dominance with guns that they obtained from the West by exporting slaves, cattle, gold and ivory. Because the northern Ovambo floodplain polities were well supplied with firearms, it took over two decades of heavy fighting—in which Western military forces suffered a series of crushing defeats—before the inhabitants of the region were subjugated and disarmed. Firearms mostly were imports, but African blacksmiths repaired and sometimes even manufactured them. Twemuna, a famous Ovambo floodplain blacksmith, reputedly even forged a breechblock for a captured Portuguese cannon, and restored the cannon to working order. Moreover,

[11] Herskovits, "The Cattle Complex in East Africa"; I. Scoones, "Range Management Science and Policy: Politics, Polemics and Pasture in Southern Africa", and W. Beinart, "Soil Erosion, Animals, and Pasture over the Longer Term: Environmental Destruction in Southern Africa", pp. 34–53 and 54–72 respectively. See also Isenberg, *The Destruction of the Bison*, and Jha, *The Myth of the Holy Cow*.

[12] Diamond, *Guns, Germs, and Steel.*

the floodplain blacksmiths' steel hoes, axes and blades were regarded as far superior to Western iron imports until well into the colonial era. Although Western industrially produced tools were available since well before World War I, local blacksmiths held their own until World War II.

Whereas the global models outlined above highlight *causes* of environmental change, the assessment of the *trajectory* (degradation, stability or improvement) and the *outcome* of such change varies according to one of the three dominant paradigms mentioned above: the modernization paradigm, the declinist paradigm or the inclinist paradigm.

The modernization paradigm

Works employing the modernization paradigm identified Western science, modern Westerners, and the species they had domesticated or adopted as the means and objectives for a state-controlled and state-exploited environment. Although his intent is to illuminate why the West colonized America, Asia and Africa rather than to celebrate the global dominance of Western modernity or Western science, Diamond's path-breaking analysis lies squarely within the modernization paradigm. Diamond identifies the early European adoption of domesticates from elsewhere—their dissemination facilitated by geo-environmental conditions—as ultimately providing Europeans with the technological and biological cutting-edge to conquer the world.[13]

If they raised environmental concerns at all, modernizers were confident that science and technology could remedy any problems that might arise, and, moreover, they judged a measure of accompanying environmental degradation to be an acceptable price for progress. For example, in Zimbabwe, the colonial-era authorities—otherwise strong proponents of game conservation—exterminated large numbers of wild animals to control tsetse fly infestation and to protect the development of white commercial cattle ranching.[14]

---

[13] Ibid. Diamond's emphasis on how a linear process of domestication enables human domination over Nature (i.e., civilization or Culture) is similar to that of, for example Sauer and Goudsblom. See Sauer, *Seeds, Spades, Hearths, and Herds*, and Goudsblom, *Fire and Civilisation*.

[14] Mutwira, "A Question of Condoning Game Slaughter".

The main objective of conservation was to prevent the irrational and wasteful use of 'natural' resources and to protect wildlife and forest resources from 'primitive' Western and non-Western farmers and pastoralists.[15] In the 1930s, the British colonial administrations in Africa became increasingly convinced of the necessity of direct intervention in how African subjects used the land.[16] Colonial officials and experts viewed 'the natives' as potential sources of pollution and disease, in addition to perceiving them as abusing or underutilizing the land. As a consequence, the officials believed that the local indigenous population should not have any rights whatsoever to lands that were not actively inhabited or cultivated. The characterization legitimized the practice of taking over as state land vast expanses of fallow, pasture lands and forests, as well as hunting and gathering grounds.[17] Although colonial officials initially regarded select indigenous peoples simply as part and parcel of Nature (e.g., as Stone Age hunters and gatherers) and consequently preserved them in the newly established reserves and parks, by the 1950s, the officials had removed the last groups of local residents from the conservation areas.[18]

To the modern colonial and postcolonial state, forests and trees especially were highly valuable economic resources that should be managed and exploited by professional foresters under the aegis of scientific forestry.[19] Tropical rainforests were valuable because they were a source of hardwoods.[20] In contrast, woodlands without desirable timber stands were viewed as wastelands that could and should

[15] See MacKenzie, *Imperialism and the Natural World* and *The Empire of Nature*; Anderson and Grove, *Conservation in Africa*, esp. pp. 1–12; Grove, *Green Imperialism*; Carruthers, *The Kruger National Park*.

[16] Anderson, "Depression, Dust Bowl, Demography, and Drought"; Berry, *No Condition Is Permanent*, pp. 46–54.

[17] Cronon, *Changes in the Land*, p. 53. On land takeovers, see M. Colchester, "Forest Peoples and Sustainability". On the view of Africans as sources of disease, see Farley, *Bilharzia*, pp. 13–20, 137–139.

[18] Konrad, "Tropical Forest Policy and Practice during the Mexican Porfirato". On removals of indigenous people from parks, see Colchester, "Forest Peoples and Sustainability", pp. 61–95; Ranger, "Whose Heritage? The Case of the Matobo National Park"; Kreike, *Re-creating Eden*, pp. 129–154; Merchant, *Reinventing Eden*, pp. 152–154.

[19] For conventional forestry see Wiersum, *Social Forestry*, pp. 27–36, 54–60; Williams, *Deforesting the Earth*, pp. 145–168, 242–275, 383–419; Guha, *The Unquiet Woods*, pp. 35–61; Peluso, *Rich Forests*, pp. 44–160.

[20] Tomlinson and Zimmermann, *Tropical Trees as Living Systems* focuses on the tropical rainforest.

be transformed into agricultural lands, for example, for the scientific production of sugarcane, cotton, cocoa, tea, coffee or other market crops.[21] In practice, however, colonial and postcolonial states frequently lacked the coherence, the capacity or the will to enforce their own conservation regulations or rationally to exploit the forest and other environmental resources. This was especially the case when colonial authorities met fierce resistance from populations that depended on forest access.[22]

The declinist paradigm

Some authors have emphasized continuity between the modernization and declinist paradigms: both highlight the danger of environmental decline.[23] The declinist paradigm, however, differs from the modernization paradigm in that it identifies (Western) modernity itself as the major cause of environmental decline.[24] Even the neo-Malthusian population bomb ultimately can be understood as having been caused by modern science: Western medicine brought mortality rates down so radically that population growth soon outpaced food production. Many historians who focused on environmental and/or agricultural change in the non-Western world have written from a declinist perspective. Often, declinists explicitly or implicitly portray precontact non-Western environments as being suspended in a state of (pristine) Nature, and precontact societies as living in harmony with Nature. Declinists argue that the modern Western economy (including capitalism, market forces and the resulting commodification of environmental resources and labor) caused overexploitation (of timber or such game animals as elephant, tiger, beaver or bison) or the diversion of precious land and labor away from food production and local resource

---

[21] See Budowski, "Perceptions of Deforestation in Tropical America", p. 1; and Tucker, "The Depletion of India's Forests under British Imperialism"; Kajembe, *Indigenous Management Systems as a Basis for Community Forestry in Tanzania*, p. 10.

[22] See D. Anderson, "Managing the Forest: The Conservation History of Lembus, Kenya"; Guha, *The Unquiet Woods*, and Peluso, *Rich Forests*; MacKenzie, "Experts and Amateurs".

[23] Fairhead and Leach, *Reframing Deforestation*, pp. 172–173; and Peluso, *Rich Forests*, pp. 44–160.

[24] See, for example, Worster, "Introduction", in *Ends of the Earth*, pp. 4–5; Pyne, *World Fire*.

management, resulting in environmental and general collapse.[25] The introduction of commercial crops or livestock also led to the clearing of forest and bush land. Some of the crops, for example, coffee and cotton, caused soil erosion.[26] Colonizers also introduced modern agriculture through large-scale commercial plantations for crops and trees, and, where lands were suitable for European settlement, through imported white farmers. Colonial administrations typically allocated prime agricultural lands to white settlers or metropolitan companies, transforming the local populations into squatters or removing them to marginal lands.[27] A related argument stressed structural imbalances in access to land and other resources as the underlying cause for deforestation: a small elite that controlled the arable land pushed poor, landless farmers into the forest wildernesses.[28]

A political ecology focus within the declinist perspective emphasizes how the modern colonial and postcolonial states sought to control—especially through conservation—not only Nature but also how the local population used and managed natural resources. Colonial administrators proclaimed forest as reserves to facilitate scientific exploitation; gazetted game reserves and national parks to protect wildlife; brought upper water catchments under government stewardship; and imposed draconian punishment to suppress indigenous burning regimes.[29] Although these measures proved difficult to enforce, they nevertheless restricted local populations' access to important environmental resources (e.g., game meat, forest products and grazing) and led to the erosion of indigenous environmental resource management. In East Africa, indigenous practices that previously had contained the impact and the spread of the trypanosomiases-

[25] See, for example, Palmer and Parsons, eds., *The Roots of Rural Poverty in Central and Southern Africa*; Pyne, *Vestal Fire*; Marks, *Tigers, Rice, and Salt*, pp. 38–40; Walker, *The Conquest of Ainu Lands*; Dean, *With Broadax and Firebrand.*

[26] See, for example, Geertz, *Agricultural Involution*; Stein, *Vassouras*; Isaacman and Roberts, *Cotton, Colonialism, and Social History in Sub-Saharan Africa.*

[27] See, for example, Beinart, Delius and Trapido, *Putting a Plough to the Ground*; Bundy, *Rise and Fall of the South African Peasantry*; Arnold, *The Problem of Nature*, pp. 119–168; Dunlap, *Nature and the English Diaspora.*

[28] See Colchester and Lohmann, *The Struggle for Land*, pp. 1–60, 99–163. On land conflict, see, for example, Durham, *Scarcity and Survival in Central America.*

[29] See Anderson and Grove, *Conservation in Africa*, pp. 1–39; Grove, *Green Imperialism*; Beinart, "Soil Erosion, Conservationism, and Ideas about Development"; Pyne, *Vestal Fire*; Guha, *The Unquiet Woods*; Peluso, *Rich Forests.*

carrying tsetse fly in Africa withered away.[30] The introduction of soil conservation projects offers a continent-wide example. During the 1930s, colonial administrations, fearing the collapse of African food production systems under the strain of environmental change and population pressure that coincidentally largely was caused by economic, political and conservation colonial policies, introduced terracing and contour plowing throughout rural Africa. Given the required extra labor demands on the local population, however, these projects often exacerbated matters, although the full weight of such policies was only felt after World War II.[31]

Biological imperialism offers a third prism through which to consider declinist environmental change. The introduction of new animals, plants and microbes or the selective favoring of indigenous species unleashed such pests and plagues as, for example, smallpox, yellow fever and sheep in the Americas, rinderpest and lungsickness in Africa, and rabbits in Australia. Some authors have emphasized that colonialism, or, more recently globalization, multiplied the impact of invading and indigenous microbes because it weakened or destroyed pre-existing environmental management arrangements.[32] Often, as is the case in the modernization paradigm, declinists depicted the scenario in terms of a precontact ecological balance.[33]

Although declinist analysis identifies modernity as the main culprit of environmental destruction, in practice, conservationist intervention often targeted indigenous communities in an attempt to change their environmental management and use strategies. Declinists sometimes admired indigenous knowledge and technology, but regarded it as traditional and static, and thus unable to

---

[30] Kjekhus, *Ecology Control and Economic Development in East African History*; Giblin, "The Precolonial Politics of Disease Control in the Lowlands of Northeastern Tanzania". On the limits of colonial policies, see also Grove, "Colonial Conservation, Ecological Hegemony and Popular Resistance".

[31] Beinart and Bundy, *Hidden Struggles in Rural South Africa*; Showers, *Imperial Gullies*; *Journal of Southern African Studies* 15 (1989), Special Issue on Conservation in Southern Africa.

[32] See Crosby, *Ecological Imperialism*; Grinde and Johansen, *Ecocide of Native America: Environmental Destruction of Indian Lands and Peoples*; Fenn, *Pox Americana*; Melville, *Plague of Sheep*; Kjekhus, *Ecology Control and Economic Development in East African History*; Giblin, "Trypanosomiasis Control in African History"; Lyons, *The Colonial*; Rolls, *They All Ran Wild*.

[33] Headrick, *Colonialism, Health and Illness in French Equatorial Africa*. Kjekhus attributes epidemic sleeping sickness to "ecological imbalances" associated with colonialism, *Ecology Control and Economic Development in East African History*, p. 166. Brooks, Webb, Johnson and Anderson

cope with the new challenges brought by the modern economy and population growth.[34] A series of devastating droughts in Africa in the 1970s and 1980s and the notion that the tropical rainforests of Africa, Latin America and Southeast Asia constituted the last and most prized remnants of pristine Nature added a sense of urgency, paving the way for radical interventions.[35]

To counter deforestation, Western experts introduced agroforestry (trees in fields) and social forestry projects in Africa, Latin America and Asia, with the goal of facilitating the reforestation of lands outside the protected forests. Focusing attention on people and their social networks and on forests and trees outside the formally declared forests, however, largely was instrumental. Because the practice of protecting existing forests from human intrusion was considered to be a failure, foresters sought to boost forest production outside the actual forests as an alternative source for the fuel wood and other products that local populations previously had gathered in the forests.[36]

In Africa, the communal woodlot approach met with little success, an outcome that in the late 1970s and early 1980s contributed to increased attention to the role of on-farm trees and farmers in agroforestry and social forestry research and projects. Yet, this micro focus was short-lived. After farm-level projects appeared to favor men over women and the wealthy over the poor, the pendulum swung back to a macro level of analysis in the 1980s and the early 1990s. Moreover, fuel

---

and Mandala show that desiccation, drought and famine also occurred in precolonial Africa, implying that a general ecological balance did not exist. See Johnson and Anderson, *The Ecology of Survival*, for example, the chapter by Pankhurst and Johnson, "The Great Drought and Famine of 1888–92 in Northeast Africa", pp. 47–70; Brooks, *Landlords and Strangers*; Webb, *Desert*; and Mandala, *Work and Control in a Peasant Economy*, pp. 15–97.

[34] Richards noted that colonial officials discovered indigenous knowledge before World War II; during the war, however, the paradigm shifted to state-led scientific approaches, *Indigenous Agricultural Revolution*, pp. 31–40. Colchester claims that the myth of the tragedy of the commons prevented a real assessment of indigenous natural resource management systems, M. Colchester, "Forest Peoples and Sustainability", pp. 61–95. On the view of indigenous knowledge as outdated, see Le Houérou, *The Grazing Land Ecosystems of the African Sahel*, and *Browse in Africa*, pp. 485–486; Núñez and Grosjean, "Biodiversity and Human Impact During the Last 11,000 Years in North-Central Chile".

[35] On desertification, see Bassett and Crummey, *African Savannas*, pp. 15–17 and Swift, "Desertification". On shifting cultivators as deforesters, see Myers, *Deforestation Rates*, pp. 4–5, 30, 45–48; and Jepma, *Tropical Deforestation*, pp. 17–21, 104–109.

[36] On agroforestry and social forestry, see King, "The History of Agroforestry", and Nair, "Agroforestry Defined"; Hobley, *Participatory Forestry*, pp. 56, 66–81; and Wiersum, *Social Forestry*, pp. 54–81, 166–170.

wood did not emerge as a key issue for farmers.[37] Instead, multipurpose trees took center stage in agroforestry and social forestry, with an emphasis on the ability of trees, especially such 'miracle trees' as the lead tree (*Leucaena leucocephala*), to enhance and maintain soil fertility and agricultural production.[38] The interest of the state, particularly forestry departments' interventions in extra-forest agroforestry, social forestry and community forestry, partly was driven by forestry imperialism legitimated in the name of conservation and rural development.[39]

The inclinist paradigm

In the mid-1990s, Fairhead and Leach turned the declinist paradigm thesis about the direction of environmental change on its head and identified forest islands not as relics of natural or climax forest vegetation (as in a declinist reading), but as a human creation.[40] A major departure from the modernization paradigm, however, was that the inherent optimism of the inclinist paradigm derived not from a belief in Western science, but from confidence in the dynamic potential of indigenous knowledge.[41]

---

[37] Wiersum, *Social Forestry*, pp. 1, 3, 62–67; Wiersum and Persoon, "Research on Conservation and Management of Tropical Forests: Contributions from Social Sciences in the Netherlands", pp. 3–4; Leach and Mearns, *Beyond the Fuelwood Crisis*, pp. 23–40, 66–67; Schroeder, "Shady Practice". On the failure of communal woodlots, see Kerkhof, *Agroforestry in Africa*, pp. 87–111.

[38] On the exaggerated wood fuel crisis and the association of forestry with agriculture, see Leach and Mearns, *Beyond the Fuelwood Crisis*, pp. 23–40. On trees and soil fertility, see Young, *Agroforestry for Soil Management*, and Huxley, *Tropical Agroforestry*, p. 280.

[39] See, for example, J. van den Bergh, "Diverging Perceptions on the Forest: Bulu Forest Tenure and the 1994 Cameroon Forest Law"; Fairhead and Leach, *Reframing Deforestation*, p. 170. See also Guha, *The Unquiet Woods*, pp. 44–45. The forest services of Indonesia and Thailand control 74% and 40% respectively of the national territories, M. Colchester, "Forest Peoples and Sustainability", p. 75.

[40] This argument was first made in Fairhead and Leach, *Misreading the African Landscape*, pp. 55–85. Fairhead and Leach extended the argument to other West African countries in their *Reframing Deforestation*.

[41] Richards, *Indigenous Agricultural Revolution*, pp. 12, 70–72, 84–85, 128–139, 151–152, 155; Leach and Mearns, *Beyond the Fuelwood Crisis*, pp. 26–40; Fairhead and Leach, *Misreading the African Landscape*. On the dynamism of African peasants, see also Berry, *No Condition Is Permanent*, pp. 49–52; Tiffen, Mortimore and Gichuki, *More People, Less Erosion*, pp. 226–245; Mazzucato and Niemeijer, *Rethinking Soil and Water Conservation in a Changing Society*.

An important second root of inclinist revisionism stemmed from the rejection of the declinists' alarmist claims, which were based on the use of prejudicial colonial information and contemporary data that were estimates at best. In his highly influential 1989 study *Deforestation Rates*, Myers predicted that little forest would be left by the end of the twentieth century. His dire prediction is still far from reality, although deforestation continues to be a major concern. Moreover, the 1976 to 1998 deforestation statistics were based on only two sets of primary sources that were themselves estimates: an FAO / UNDP analysis that relied partly on satellite data and Myers' own study.[42] Boserup's *Conditions of Agricultural Growth*, which argues that population pressure gives rise to technical innovation and the intensification of land use, further strengthens the inclinist world view.[43]

In the inclinist paradigm, indigenous knowledge and indigenous management and use of forest resources take center stage as points of departure for research and intervention.[44] The definition of what constituted 'forest' further was expanded to include the dry forests (including the *miombo* expanses of Africa) and woodlands that support much larger populations than the rainforests.[45] Inclinists consider indigenous populations not as an environmental threat, but as a critical part of the solution.[46] Social forestry included transferring 'forest' management from

---

[42] See Leach and Mearns, *Beyond the Fuelwood Crisis*, pp. 1–9; Fairhead and Leach, *Misreading the African Landscape*, pp. 1–85, 121–136, 182–197, 237–278; McCann, *Green Land, Brown Land*, pp. 79–107; Bassett and Crummey, *African Savannas*, pp. 4–15, 24; Lehman, "Deforestation and Changing Land Use Patterns in Costa Rica", p. 67. Although all the contributors in Steen and Tucker acknowledge deforestation as an important issue, a number of them reject declinism as a straightjacket; see, for example, the chapters by Pierce (pp. 40–57), Lehman (pp. 58–76), Graham and Prendergast (pp. 102–109) and Balée (pp. 185–197). See Myers, *Deforestation Rates*, p. 4, and Williams, *Deforesting the Earth*, pp. 477–479, 453–457.

[43] Boserup, *Conditions of Agricultural Growth*; Pingali, Bigot and Binswanger, *Agricultural Mechanization*. See also Leach and Mearns, *Beyond the Fuelwood Crisis*, pp. 1, 53; Tiffen, Mortimore and Gichuki, *More People, Less Erosion*; and Siebert, "Beyond Malthus and Perverse Incentives", p. 29.

[44] Leach and Mearns, *Beyond the Fuelwood Crisis*, pp. 23–40. Franzel et al. emphasize the importance of building on Indigenous Technical Knowledge (ITK), Franzel, Cooper, Denning and Eade, eds., *Development and Agroforestry*, see especially the contributions by Denning (pp. 1–14), Haggar et al. (pp. 15–23), Weber et al. (pp. 24–34) and Wambugu et al. (pp. 107–166). See also Balée, "Indigenous History and Amazonian Biodiversity".

[45] See Westoby, *Introduction to World Forestry*, pp. 147, 169–170. On the *miombo* woodlands, see Campbell, *The Miombo in Transition*.

[46] Several chapters in Franzel and Scherr underline the importance of on-farm participatory research with farmers but stress that the scientists need to remain in control, see Franzel et al., "Methods of Assessing Agroforestry Adoption Potential", and Scherr and Franzel, "Promoting Agroforestry Technologies: Policy Lessons from On-Farm Research".

the state to local communities, although in practice, officials and scientists overwhelmingly proved incapable or unwilling to relinquish real control over conservation areas and experiments.[47] In India, for example, the state continued to set the agenda in joint state-local community forest management projects, a practice that resembled colonial indirect rule because it relied on (unpaid) 'traditional' local leaders for enforcement.[48]

Paradoxes of environmental change

The modernization, declinist and inclinist paradigms each offer important insights into the dynamics of environmental change. Because they are cast as being competing and mutually exclusive, however, the paradigms create paradoxes about the process of environmental change. The first paradox is the presence of such remnants of urban settlements as, for example the ruins of Palenque, Mexico, in pristine forest. The urban environment was and is a powerful symbol of the dominance of Culture over Nature, representing the apex of civilization to modernizers, and Nature's nadir to declinists. The urban environment also is seen to be the antithesis of wilderness in the Nature-Culture dichotomous framework that the three paradigms share.[49] The benchmark environment against which environmental change is assessed and measured is variously referred to as wilderness, Nature, pristine Nature, State of Nature / Natural State, precontact environment (indigenous Edens or people-Nature balances) or vegetation climax.[50] The defin-

---

[47] On indigenous farmers' participation and its limits, see Leach and Mearns, *Beyond the Fuelwood Crisis*, pp. 230–231; Denning, "Realising the Potential of Agroforestry: Integrating Research and Development to Achieve Greater Impact"; Haggar et al., "Participatory Design of Agroforestry Systems: Developing Farmer Participatory Research Methods in Mexico"; Weber et al., "Participatory Domestication of Agroforestry Trees: An Example from the Peruvian Amazon"; and Wambugu et al., "Scaling Up the Use of Fodder Shrubs in Central Kenya".

[48] Sundan, "Unpacking the 'Joint' in Joint Forest Management". See also Peluso, *Rich Forests*, pp. 124–165; Hobley, *Participatory Forestry*, pp. 59–60, 80, 130, 139–157, 191–193, 244, 251, 259–260; Fairhead and Leach, *Reframing Deforestation*, pp. 192–193.

[49] The classic study on the concept of wilderness is Nash, *Wilderness in the American Mind*. Cronon and White argue for a Nature-Culture (urban-rural and wild-domesticated) continuum, see Cronon, *Nature's Metropolis*, pp. 17–19; and White, *The Organic Machine*, pp. 105–109.

[50] Blaikie and Brookfield, for example, posit an Edenic point of departure; see *Land Degradation and Society*, p. xx. On discomfort with the climax concept, see Longman and Jeník, *Tropical Forest and Its Environment*, pp. 13–14, 20–21, 25; Kozlowski, Kramer and Pallardy, *The Physiological*

ing characteristic essentially is the same: the absence of human action in shaping the environment. As humans affected the environment, it changed from its pre-human contact state. The closer the human communities are perceived to be to the 'Natural State', the less they are thought to change their environment (either for the worse or for the better, depending on the paradigm). For example, until recently, conventional wisdom maintained that indigenous people who live by Nature as hunter-gatherers do not shape their environment. The impact of indigenous peoples on the environment at the turn of the twentieth century became hotly debated.[51]

Indeed, the very idea of assessing and measuring environmental change along a Nature-Culture gradient with Nature as the point of departure and Culture as the outcome created a paradox. The principal remaining vestiges of unspoiled Nature, that is, the forest regions of Central and South America and Southeast Asia, as well as the proverbial last Wilderness Continent, Africa, contain such 'lost cities' as, for example, Palenque in Mexico's rainforest and Thulamela in South Africa's Kruger National Park.[52]

Neither Palenque nor Thulamela were exceptional or isolated anomalies in an otherwise pristine wilderness. Thulamela was associated with Great Zimbabwe, which stood at the center of a trade network that linked it to a global hinterland that stretched through much of southern Africa and across the Indian Ocean to India, Southeast Asia and China.[53] For comparison, modern Vancouver's hinterland is 318 times the actual size of the city, with the city and its population using the biophysical output of 3.6 million hectares scattered across the entire globe. Chicago's urban growth similarly consumed the resources of an enormous

---

Ecology of Woody Plants, p. 100; Pimentel, Westra and Noss, *Ecological Integrity*; and L. Westra et al., "Ecological Integrity and the Aims of the Global Integrity Project", ibid., pp. 19–41. For a critical overview, see Fairhead and Leach, *Reframing Deforestation*, pp. 10–11, 20, 24, 164–166.

51 For hunter-gatherers as living by Nature, see Sahlins, *Stone Age Economics*, p. 27; and Lee, "What Hunters Do for a Living". For critiques of the concept of a premodern human-nature balance, see Krech, *The Ecological Indian: Myth and History*; Isenberg, *The Destruction of the Bison*; Wingard, "Interactions between Demographic Processes and Soil Resources"; and MacLeod, "Exploitation of Natural Resources in Colonial Central America".

52 On Palenque, see Stuart and Stuart, *Lost Kingdoms of the Maya*, pp. 19, 31; and Perera and Bruce, *The Last Lords of Palenque*, pp. 10–26. On Thulamela, see Davidson, "Museums and the Reshaping of Memory", pp. 150–151. On Africa as the last wilderness, see Adams and McShane, *The Myth of Wild Africa: Conservation without Illusion*, chap. 1.

53 Hall, *The Changing Past*, pp. 91–116.

hinterland, dramatically transforming the city's environment in the process.[54] The lost cities in the African, the Latin American and the Southeast Asian wilderness similarly must have left extensive environmental footprints. Even before the twentieth century, the primordial forest and woodland of much of the Americas, Southeast Asia and Africa were shaped heavily by human use. The forests that hide the Maya ruins might be no more than four hundred years old and they differ in composition from the pre-Mayan era woody vegetation. The pristine rainforest of Suriname in the seventeenth and eighteenth centuries was the locus of a thriving plantation system that collapsed with the abolition of slavery. Today's forests in the northeastern United States grew on abandoned agricultural lands. The jungles of Kalimantan cover the ruins of mighty Srwijaya, which thrived from the sixth to the fourteenth century AD. The forest 'wilderness' of southeastern Borneo in the seventeenth and eighteenth centuries was not only extensively used for shifting cultivation and permanent agriculture, but also for commercial agriculture.[55] Africa's 'wild' landscapes similarly are arguably human creations: for example, the West African forest islands that Fairhead and Leach studied were human-made and the extensive *miombo* woodlands of eastern and southern Africa have been modified by human use. Indeed, the very idea of 'Wild Africa' is a myth.[56]

---

[54] Meggers, "Natural Versus Anthropogenic Sources of Amazonian Biodiversity"; and Cronon, *Nature's Metropolis*, pp. 17–19.

[55] On the Maya, see Leyden, Brenner, Whitmore, Curtis, Piperno and Dahlin, "A Record of Long- and Short-Term Variation from Northwest Yucatán", and Wingard, "Interactions between Demographic Processes and Soil Resources in the Copán Valley, Honduras." For similar arguments regarding northern Mexico and the Amazon, see Alcorn, "Huastec Noncrop Resource Management", and Becker and León, "Indigenous Forest Management in the Bolivian Amazon". On Suriname, see Boomgaard, "Exploitation and Management of the Surinam Forests". On the United States, see McShea and Healy, eds., *Oak Forest Ecosystems*, pp. 4–5, 13–33, 34–45, 46–59 and 60–79. On Srwijaya, see McNeely, "Foreword", in Sponsel, Headland and Baily, *Tropical Deforestation*, pp. xv–xvii. On Borneo, Knapen, *Forests of Fortune?* pp. 189–281. See also Rietbergen, *The Earthscan Reader on Tropical Forestry*, pp. 1–2; Boyce, *Landscape Forestry*, p. vii; Sponsel, Headland and Baily, "Anthropological Perspectives on the Causes, Consequences, and Solutions of Deforestation", pp. 7–8; Longman and Jeník, *Tropical Forest and Its Environment*, pp. 13–14, 24 and 27.

[56] On Africa, see Adams and McShane, *The Myth of Wild Africa*, pp. 1–13; McCann, *Green Land, Brown Land*, p. 2; Sheperd, Shanks and Hobley, "Management of Tropical and Subtropical Dry Forests", pp. 107 and 112; Fairhead and Leach, *Reframing Deforestation*; Berry, *Cocoa, Custom, and Socio-Economic Change in Rural Western Nigeria*, p. 66; Webb, *Desert Frontier*, p. 3; Campbell, *The Miombo*, pp. 1–3; Kreike, *Re-creating Eden*, chaps. 1–4; Ford, *The Role of Trypanosomiases in African Ecology*; Kjekhus, *Ecology Control and Economic Development in East African History*.

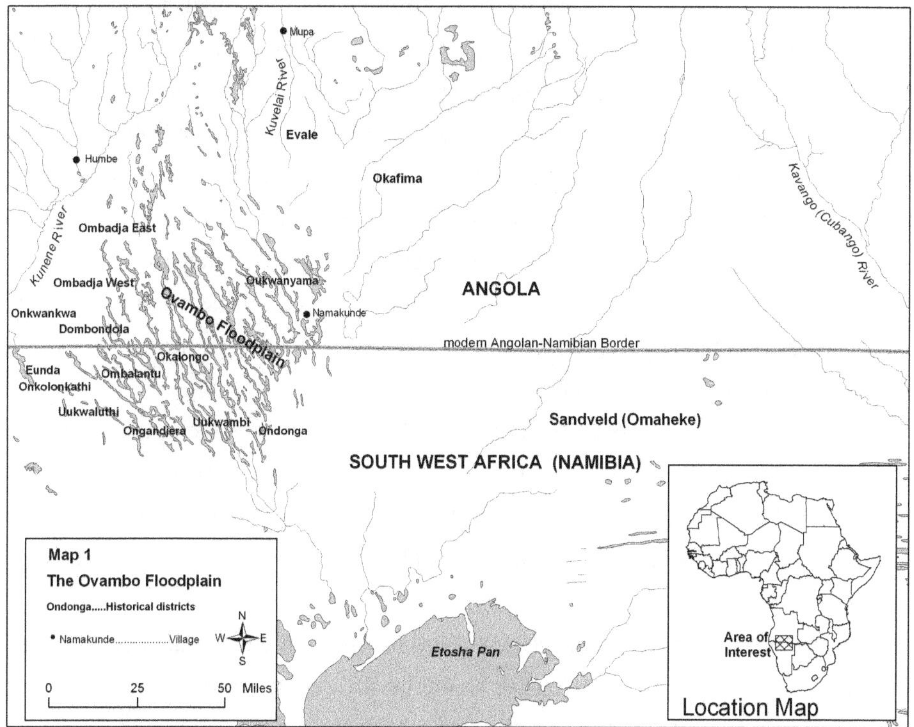

Whereas the Palenque Paradox problematizes unilinearity and static outcomes, the Ovambo Paradox suggests that deforestation and reforestation may occur simultaneously and that environmental change cannot be understood as a singular process. Violent Portuguese conquest of the northern Ovambo floodplain (in modern southern Angola) during the first two decades of the twentieth century caused massive population displacement into the uninhabited wilderness area of the middle Ovambo floodplain and the Sandveld to its east (in modern northern Namibia). As the refugees settled the wilderness areas, they deforested land in order to construct farms, fields and villages. The impact of the refugee resettlement on the woody vegetation of the area was particularly dramatic in the 1920s and 1930s.

Paradoxically, as the deforestation of the wilderness areas in northern Namibia progressed, a process of reforestation followed in its wake. The refugee-settlers and their descendants propagated and often introduced the majestic fruit trees that during the 1960s, 1970s and 1980s shaded many a farm in the middle floodplain in Namibia. Deforestation and reforestation, however, was neither cyclical (as in a natural return to a vegetation climax) nor discretely sequential; rather, multiple contradictory subprocesses of deforestation and reforestation occurred simultaneously. For example, a single village consisted of both older and more recently arrived households. Some of the latter had only just cleared their plots of woody vegetation, while some of the former had done so several decades previously, and in the meantime had reforested their plots. Thus, overall, north-central Namibia saw dramatic environmental changes in less than a century: many areas were heavily deforested and reforested, revealing multitrajectory and contradictory environmental changes.[57]

Contradictions and ambiguity in the record of environmental change have been noted elsewhere.[58] Beyond the recognition that the outcome of the process may be evaluated differently by different stakeholders, however, such acknowledgment has not led to questioning the *homogeneity* of the process of environmental change itself.[59]

The differentiation in the processes of environmental change also is obscured by a fixation on the outcome rather than the process itself. Huxley noted that "[e]cologists often study the *outcome* of plant-plant interactions in terms of changes in species number. Unfortunately, because the *processes* involved are extremely complex, less is known about these in most cases".[60] Huxley's observation is equally relevant to how environmental change as a whole has been studied using the modernization, declinist and inclinist paradigms: late twentieth-century research emphasized the outcome of Human-Nature interactions (degradation, stabilization or improvement) more than the processes themselves.[61] For example, a comparison of two photographs or two sets of aerial photography / satellite images from different times can show differences in vegetation cover and facilitate an assessment about, for example, deforestation or reforestation, but the compar-

---

[57] Ibid., pp. 137–180; Kreike, "Hidden Fruits".

[58] Moore and Vaughan, *Cutting Down Trees*; Fairhead and Leach, *Misreading the African Landscape*; Meggers, "Natural Versus Anthropogenic Sources of Amazonian Biodiversity: The Continuing Quest for El Dorado", p. 89; Gibson, McKean and Ostrom, "Explaining Deforestation", p. 2; Schama, *Landscape and Memory*, pp. 9–10.

[59] Blaikie and Brookfield, *Land Degradation and Society*, pp. 4–7, 14–16.

[60] Huxley, *Tropical Agroforestry*, p. 135.

[61] Williams, *Deforesting the Earth*, p. 237.

ison provides no information about the process of change itself. And, even if no substantial change in vegetation cover can be detected between the two measuring points, it is possible that the actual composition of the vegetation itself has changed dramatically.[62]

Such issues may be more acute in Africa than elsewhere, not only because deforestation data (and other environmental statistics) for the continent are largely nonexistent or questionable, but also because more of the environmental change is caused by individuals and households for their own benefit than is the case in Latin America, for example, or in Southeast Asia.[63] In Latin America, especially in the Amazon, and in Southeast Asia, in particular in Indonesia, state and commercial interests play a much more direct role in encouraging deforestation through colonization schemes, timber exploitation, plantation agriculture or ranching. State and commercial clearings are larger and more concentrated and therefore leave a much more distinct environmental footprint that can be detected in aerial photography and satellite imagery. In addition, state and commercial enterprises produce more information about their activities because they often are controversial. In Africa, forest settlement is more spontaneous, and small-scale individual clearings, even if they are numerous, are virtually impossible to detect on Landsat images, especially since selected trees and bush often are spared when farms are cleared. Such images therefore, cannot identify pristine Nature or climax vegetation even if they exist. In short, the images cannot unambiguously distinguish rural cultural from natural landscapes.[64] The analysis that follows seeks to address the challenge outlined above within the context of exploring global paradigms and local paradoxes through the case of north-central Namibia.

---

62 Mazzucato and Niemeijer, *Rethinking Soil*, pp. 125–127.

63 Williams, *Deforesting the Earth*, pp. 401–406; and Gibson, McKean and Ostrom, "Explaining Deforestation".

64 See Fairhead and Leach, *Reframing Deforestation*, pp. 8–9; Balée, "Indigenous History and Amazonian Biodiversity", pp. 187–188; Vandermeer, "The Human Niche and Rain Forest Preservation in Southern Central America"; Williams, *Deforesting the Earth*, p. 477. On Southeast Asia and Latin America versus Africa, see Colchester, "Colonizing the Rainforests", pp. 5–9.

2

Tree castles and population bombs

Population pressure has been identified as the major force for environmental change in the twentieth century.[1] Yet, while macrolevel analysis of the interaction between human populations and the environment demonstrates that population dynamics relate to environmental change, the correlation does not always originate from a direct causal relationship. Moreover, the relationship between population pressure and environmental change and the outcomes of change is not necessarily linear. Rather the impact of population density on a forest environment is ambiguous and multifaceted. *Where* and *how* people impacted on local environmental resources was as important as *how many* people affected the environment of north-central Namibia.

Malthus argued that population increased at a far greater rate than food production, and neo-Malthusian analysis identifies population growth as the principal cause of deforestation in Africa, Asia and Latin America. Boserup and others, on the other hand, stress that population growth can have the opposite effect because intensification and technological innovation can permit the same resource base to support a larger population without environmental degra-

---

[1] Myers, *Deforestation Rates*, pp. 20–23, 45–47; Williams, *Deforesting the Earth*, pp. 168–209, 334–379, 460–466.

dation.[2] Both approaches portray 'population' and 'forest' as undifferentiated and organic entities. Moreover, the relationship between the two variables is depicted as being a mechanical, linear, one-way and unequal interaction, i.e., human populations are dominant and act upon the forest.[3] The population pressure model to some extent approximates cultural determinism, as opposed to environmental determinism. The underlying causes of population growth, however, sometimes are couched in terms of biological determinism; for example, in *The Population Bomb* Ehrlich writes: "our urge to reproduce is hopelessly entwined with our other urges".[4] In essence, while humans (or Culture) are advanced as the cause of environmental change, they are not really considered to be independent agents; rather, they are hostages to biological urges.

Malthusian and Boserupian explanations are particularly influential in the case of modern Africa because the continent has the highest rates of natural population increase. Two issues, however, complicate matters. First, a number of the African countries that are listed amongst those with the highest deforestation rates, including Gabon, Congo (Brazzaville) and the Democratic Republic of Congo, are underpopulated.[5] Second, research suggests that Africa's population began to grow only in the 1940s or 1950s, although environmental degradation related to population growth, notably deforestation and soil erosion, became major concerns in the late 1920s and the 1930s.[6] Population movements, however, led to the relative redistribution of the existing population, with concentrations of

---

[2] Malthus, *An Essay on the Principle of Population*. See also Ehrlich, *The Population Bomb*; Ehrlich and Ehrlich, *The Population Explosion*; Cleaver and Schreiber, *Reversing the Spiral*. On Boserup-inspired approaches see Boserup, *The Conditions of Agricultural Growth*; Pingali, Bigot and Binswanger, *Agricultural Mechanization and the Evolution of Farming Systems in Sub-Saharan Africa*; Okafor and Fernandes, "The Compound Farms of South-Eastern Nigeria: A Predominant Agroforestry Homegarden System with Crops and Small Livestock"; Tiffen, Mortimore and Gichuki, *More People, Less Erosion*; Quisumbing and Otsuka, *Land, Trees and Women*, pp. 43–79; and Siebert, "Beyond Malthus and Perverse Incentives", pp. 19–21.

[3] For critiques of the population pressure models, see, for example, Cordell and Gregory, *African Population and Capitalism*, pp. 14–15; J. Koponen, "Population: A Dependent Variable", and G. Maddox, "Environment and Population Growth in Ugogo, Central Tanzania"; Fairhead and Leach, *Reframing Deforestation*, pp. 13, 178; Mazzucato and Niemeijer, *Rethinking Soil and Water Conservation*, pp. 124–164.

[4] Ehrlich, *The Population Bomb*, pp. 31–32. For a critique, see Koponen, "Population: A Dependent Variable", and Maddox, "Environment and Population Growth".

[5] Myers, *Deforestation Rates*, pp. 20–23, 45–47; Westoby, *Introduction to World Forestry*, p. 109.

[6] Koponen, "Population: A Dependent Variable", and Maddox, "Environment and Population Growth"; Notkola and Siiskonen, *Fertility, Mortality and Migration in Subsaharan Africa*, chap. 9; Headrick, *Colonialism, Health and Illness in French Equatorial Africa*, pp. 89, 194–195, 385–394.

specific groups of people and subsequent population pressure in some areas, and depopulation and decreasing population pressure in others. Thus, until the 1940s or 1950s, population movement in Africa may have been a more critical variable than population growth, and indeed migrations continue to play a major role in the population dynamics of modern Africa and consequently in environmental changes.[7]

In pre-World War II north-central Namibia's Ovamboland Native Reserve, environmental change was driven more by population movements than by natural population growth. Insecurity and security concerns are key to explaining why, how, where and when populations movements are associated with deforestation. In the late 1800s through the early 1920s, a general climate of insecurity caused people to concentrate in nucleated wooden fortifications—tree castles—for purposes of defense. The fortifications were extensive and elaborate, and building them required enormous amounts of wood. From the 1920s onward, improved political security allowed people to fan out from population centers into the surrounding wilderness. As settlers moved into the wilderness, however, they faced a new threat: wild animal populations that had rebounded from 1890s lows. To protect lives and livelihoods, the settlers retained the practice of constructing fortified homesteads, contributing to a high consumption of woody vegetation and deforestation.

Tree castles and insecurity on the eve of colonial conquest

Warfare and violence associated with the slave and ivory trades and colonialism at the turn of the twentieth century caused the inhabitants of modern southern Angola and northern Namibia to concentrate for safety and to invest significant

---

7 Iliffe highlights migrations and the colonization of new environments as the main characteristics of the history of Africa, see Iliffe, *Africa*. For examples of studies that emphasize the environmental impact of migrations, see McCann, *Green Land, Brown Land*, pp. 19–22, 96–101; Maddox, "Environment and Population Growth", esp. p. 44; Miehr, "Acacia albida and Other Multipurpose Trees"; Kessy, *Conservation and Utilization of Natural Resources in the East Usambara Forest Reserve*, p. 60; Colchester and Lohmann, *The Struggle for Land*, especially the contributions by Colchester (pp. 1–15), Lohmann (pp. 16–34), Plant (pp. 35–60), Colchester (pp. 99–138) and Monbiot (pp. 139–163); Gray, "Investing in Soil Quality", esp. pp. 73, 76–77.

resources and labor in defensive works. Woody vegetation was the main construction material for elaborate fortresses shaped in the form of circular labyrinth palisades or enclosures in the Ovambo floodplain that straddled the Angolan-Namibian colonial border. Communities constructed the palisades using nine-to-twelve-foot-high poles buried three feet in the soil, resulting in fortifications that were impregnable to spears, arrows and even modern small arms. In the kingdom of Okafima, in the far northeastern Ovambo floodplain, the royal fort was sufficiently large to provide shelter to all of its 1,500 inhabitants. In the southwestern part of the floodplain—including Ombalantu, Uukwaluthi and Eunda—huge baobabs served as medieval-style keeps. In Ombalantu, people constructed their homesteads in close proximity to such forts. When an attack was feared, women and children sought safety in the hollow trunk of the baobab, where water usually was stored, while livestock was driven inside the palisade around the tree. Some of the baobab castles had mud-plastered outer and inner palisades. Archers positioned themselves on platforms behind loopholes. The well-known Ombalantu baobab that is now a national monument is a good example of such a former keep. A South African official "had a doorway cut in and used the room, in which upwards [of] 50 people can stand, as a store". The construction and maintenance of the fortifications consumed large quantities of wood and labor. When the Oukwanyama King Weyulu moved his palace-fortress over a short distance in late 1895, seventy men were engaged in cutting and transporting new poles while others laid out the ground plan.[8]

Not only did kings and other notables reside in formidable tree castles, but the homesteads of the local populations also were protected by palisade enclosures which typically consisted of wooden poles. The homesteads required great effort to construct, and they contained from twenty to seventy open or closed 'huts'.

---

[8] Kreike, *Re-creating Eden*, chaps. 2–3; NAN, WAT ww17, S. Davis, "Tour of Northern Territories—Some Random Comments and Thoughts"; A 450, 4, Hahn to A.W. Hoernle, 21 Sep. 1936; RCO 4, RCO to GRN, 29 Feb. 1916; RCO 8, RCO to Sec. SWA, Ondonga, 27 Oct. 1918 and Extract from RCO's Personal Diary, 10 March 1917; NAO 104, Anderson to Hahn, 13/4/44 Johannesburg 1882, extract diary W.W Jordan copied from the *Cape Quarterly Review* 2 (1882): 519–539; AVEM, RMG 2599 C/i 19, Bernsmann, Omburo, 6 Jan. 1892, and RMG C/h 52, Speiker, Visitationsbericht der Station Namakunde, Namakunde 13–18 July 1906; Möller, *Journey in Africa*, pp. 110–112, 117; Lima, *A Campanha dos Cuamatos*, pp. 181–183; CNDIH, Avulsos, Caixa 3439, Ribeiro da Fonseca, "Relatório do reconhecimento", Cuamato, 26 Sep. 1913; OMITI 4.4.38. On Dayak village fortresses, see Knapen, *Forests of Fortune?* pp. 86–88.

Photo 1. Mwanyangapa's Baobab Castle, Ombalantu, 1917. The baobab which served as the keep is surrounded by a mud-plastered palisade (Iziko Museums, Cape Town, Dickman Collection)

In 1850s Ondonga, the first homestead that the missionary Carl Hugo Hahn encountered measured approximately sixty feet across and consisted of poles and stalks planted in the earth forming a small labyrinth. There were separate huts for each of the wives and separate compartments for livestock. In the early 1890s northern floodplain, bundled thorn branches were sometimes used to make palisades instead of poles; in Ondonga and Uukwambi in the southern floodplain, where poles and branches were in shorter supply, palisades constructed from bundles of grain stalks were common.[9] A thorn bush fence—which was as effective as barbed wire—surrounded the palisaded homestead and its fields.[10]

---

[9] Lau, *Carl Hugo Hahn Tagebücher*, vol. 4, 22 July 1857; AVEM, RMG 2599 C / i 19, Bernsmann, Omburo, 6 Jan. 1892; Möller, *Journey in Africa*, p. 126. On the great labor investment required, see, for example, AVEM, RMG C / h 52, I. Speiker, Visitationsbericht der Station Namakunde, Namakunde 13–18 July 1906; Wülfhorst, *Moses*, pp. 14–15. The homestead was called *eumbo* in Oukwanyama and *egumbo* in Ondonga.

[10] See, for example, NAN, NAO 104, Jordan diary; and Lima, *A Campanha dos Cuamatos*, pp. 136–140, 159.

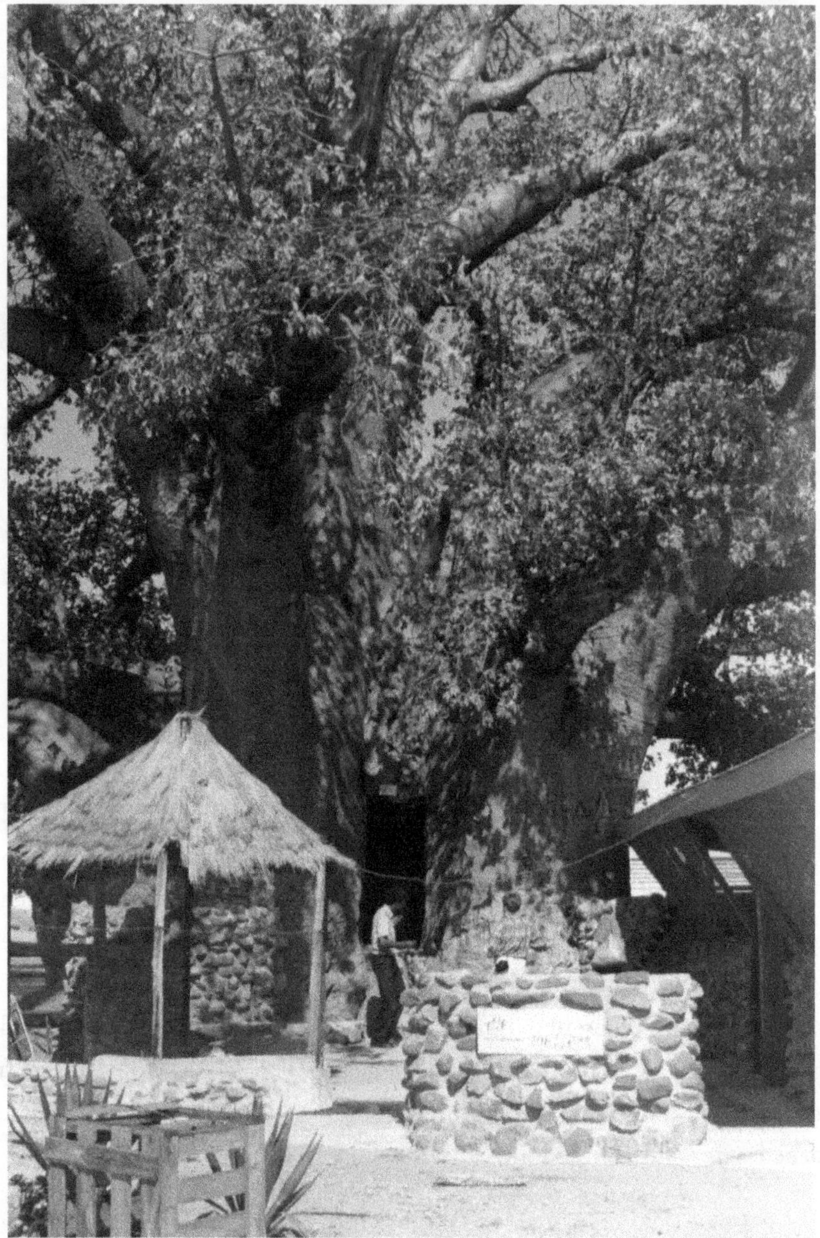

Photo 2. The Ombalantu Baobab, 1993. A fort in the 1910s, the tree was used as a colonial post office and a chapel. During the 1970s and 1980s Apartheid Wars, it served as a bar in a South African military base. The large door cut in the trunk during the early colonial era is pointed out by Joseph Hailwa, the Namibian Director of Forestry (Photo by author, 1993)

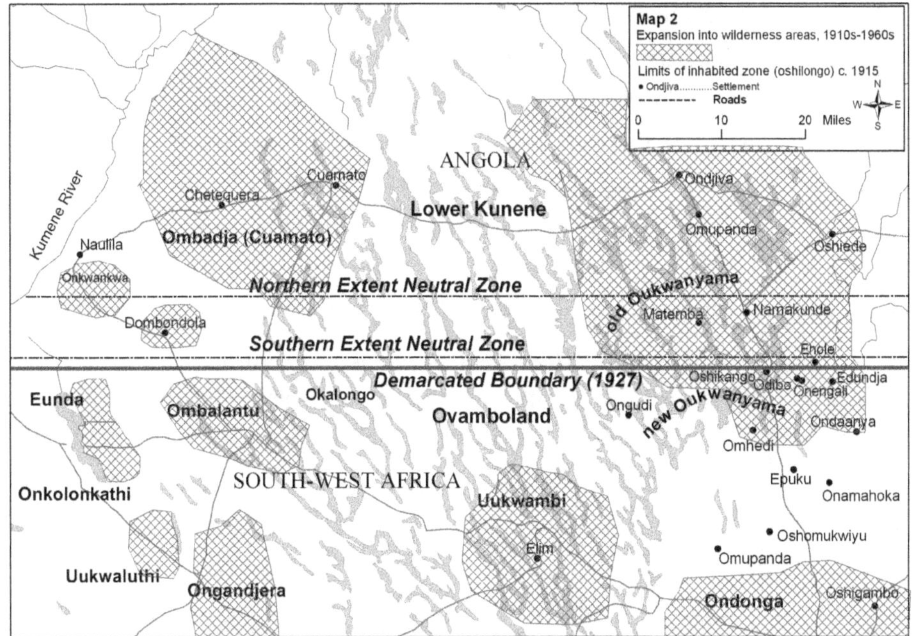

Portuguese violence and population flight into Ovamboland

Colonial conquest and pacification caused massive destruction and large-scale population displacement across Africa. In the northern Ovambo floodplain, the violence and terror caused by the Portuguese colonial conquest and 'pacification' of the Ombadjas, Evale and Okafima from 1909 to 1915 and of Oukwanyama from 1915 to 1917 made tens of thousands of the inhabitants flee south into the South African-controlled middle and southern floodplain. Many refugees from the northern floodplain initially resettled in the Neutral Zone, a disputed territory that between 1915 and 1927 was co-administered by the South African and the Portuguese colonial powers until it was ceded to Portugal. The cession triggered yet another large-scale migration to the middle floodplain area south of the new colonial boundary. It was then that many refugees who had settled in the Neutral Zone moved into the South African-occupied part of the Ovambo floodplain which became known as Ovamboland.[11]

[11] Kreike, *Re-creating Eden*, chaps. 1–4.

In 1916, only a few of the refugees had managed to construct fortress-like homesteads and the impact of the refugee movement on the vegetation was as yet marginal: "this country is a vast flat, covered with more or less thick bush, grass and tall antheaps (...). there are also large dry uninhabited stretches (...). Except for [the Oukwanyama king] Mandume's and some principal headmen's stockaded kraals the natives live separated in unimportant little huts scattered about in the bush".[12]

By 1933, however, a substantial share of Ovamboland's population consisted of refugees from the northern floodplain polities of Oukwanyama, the Ombadjas, Evale and Okafima. Refugees from the northern floodplain escaped to Ondonga and Ongandjera in the southern floodplain, and also sought sanctuary in the wilderness of the middle floodplain. Colonial Namibia's tribal district of Oukwanyama, which was nonexistent in 1915, in 1933 counted 41,215 inhabitants, or 38 % of Ovamboland's total population of 107,861. By 1938, the district's population had increased to 52,580.[13]

The impact of flight on the population of what became north-central Namibia was still visible in the 1991 census data. Of the survivors of the generation born before 1917, 18 % claimed to be foreign born. Of those born between 1917 and 1926, 15 % were foreign born, a figure that decreased to 11 % for people born between 1927 and 1936; 9 % for those born between 1937 and 1946; 8 % for those born between 1947 and 1976; and less than 5 % for those born after 1976.[14]

Internal migration in South Africa's Ovamboland

A second settlement movement into the wilderness was the result of internal migrations within Ovamboland. By the mid-1920s, the southern and middle floodplains had become much less violent spaces and people no longer felt compelled to seek safety in numbers. As a result, settlers struck out into the wilderness

---

[12] NAN, UNG, UA 1, RCO to Deputy Secretary for the Protectorate, Ondonga [Ondangwa], 31 May 1916. Interviewees also recall that their farms initially were small, see, for example, Paulus Nandenga, interview by author, Oshomukwiyu, 28 April 1993.

[13] Kreike, *Re-Creating Eden*, chap. 4.

[14] Derived from Republic of Namibia, *1991 Census*, vol. 1, table A03.

areas that had separated the old kingdoms. Again, this was a phenomenon that was not unique to Ovamboland. During the 1920s, 1930s and 1940s, individuals and groups of Africans throughout the continent fanned out from defensible sites with dense settlement to occupy wilderness expanses.[15]

Between 1915 and 1950, migrants entirely transformed the vast wilderness of the middle floodplain. By 1928, bush dominated the district, although the flood channels themselves mostly were bare of woody vegetation.[16] The large majority of the area's modern villages date to the post-1915 era and woody vegetation to construct farms and fences was abundant during the 1920s through the 1940s. In the 1940s, such middle floodplain villages as Omupanda, Oshapwa and Oshomukwiyu, for example, contained only a few households, but Odimbo, seven miles north of Oshomukwiyu, already was densely settled. Ombadja refugees founded a small cluster of villages during the 1920s further west.[17] Refugees from the northern floodplain also settled throughout the southern floodplain, including in Ondonga's eastern Oshigambo area.[18]

---

[15] Iliffe, *A Modern History of Tanganyika*, pp. 315–317. For other examples of such population dispersals, see Mandala, *Work and Control in a Peasant Economy*, pp. 95–96; McCann, *Green Land, Brown Land*, pp. 147–156; van Beek and Banga, "The Dogon and Their Trees"; Miehr, "Acacia albida and other Multipurpose Trees"; Maack, "'We Don't Want Terraces!'", esp. p. 155; Richards, *Indigenous Agricultural Revolution*, p. 54; Mazzucato and Niemeijer, *Rethinking Soil and Water Conservation*, p. 82. Similar processes marked Southeast Borneo and the Jimi Valley of Papua New Guinea. Knapen, *Forests of Fortune?* pp. 252–253; and F. van Helden, "Resource Dynamics, Livelihood and Social Change on the Forest Fringe", p. 87.

[16] NAN, KAB 1, Volkmann, 30 Oct. 1928, Report on the Agricultural and Political Conditions at the Angola Boundary; RCO 9, RCO to Sec. Protectorate, 18 Feb. 1917; RCO 10, RCO and Hahn, Preliminary Memo re Ovamboland and Chief Mandume [1916].

[17] Kreike, *Re-creating Eden*, chaps. 2–4; NAN, NAO 18, Monthly Reports Ovamboland, June – July 1927; A450, 9, Newspaper clipping of an article by G.W. Tobias [1925]; NAO 16, UGR to O/C NAO, Namakunde, 28 Feb. 1929; NAO 44, NCO to Sec. SWA, Ondangwa 20 April 1931; NAO 10, O/C NAO to NCO, Oshikango, 19 June 1935; interviews by author: Mateus Nangobe, Omupanda, 24 May 1993; Joseph Shuya, Oshapwa, 23 June 1993; Paulus Nandenga, Oshomukwiyu, 28 April 1993 and Paulus Wanakashimba, Odimbo, 10–11 Feb. 1993.

[18] CNDIH, Caixa 4130, Governo do Distrito 31-20 1, Relatorio sobre a occupação, 18 June 1909, and Avulsos Caixa 3439, Ribeiro da Fonseca, Relatorio do reconhecimento, Cuamato, 26 Sep. 1913; NAN, RCO 4, RCO to British Resident Namakunde, 13 Dec. 1915 and to GRN [Ondangwa], 29 Feb. 1916; RCO 5, Pearson to RCO, T'Chipa, 1 March 1920; A450, 7, Native Affairs Report Ovamboland 1924, and vol. 10, document marked in right top corner "By Shovala 17/9/35"; NAO 18, Monthly Report Ovamboland, Sep. 1926; NAO 106, Diaries NCO, Diary 1935–1938, 9 Sep. 1935 and 30 June 1936; NAO 9, UGR to O/C NAO, Namakunde, 28 Aug. 192 and Bourquin, memo re Headman Filemon Shipena of Elope-Ondonga, [22 Jan. 1942]; NAO 51, Shimoshili to Master Nakale [Eedes], Olupembana, Uukwambi, 13 Jan. and 2 June 1948; Nauyoma Kapeke [to NCO], Uukwambi [received Ondangwa, 3 April 1948]; NAO 51, NCO to CNC, Ondangwa,

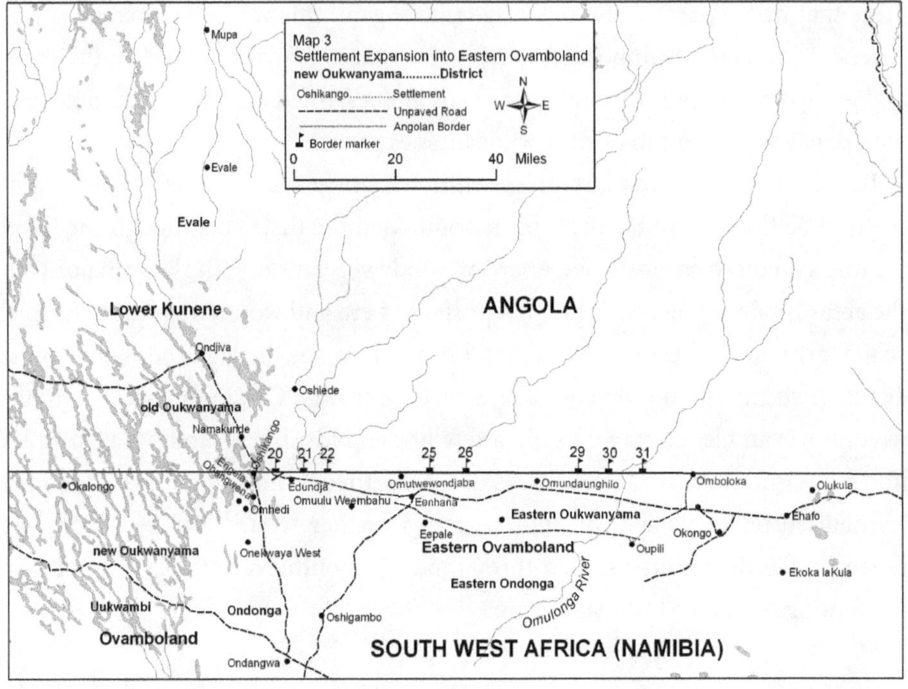

In March 1917, the inhabitants of Ombalantu huddled in farms concentrated around baobab forts. In 1918, the formidable Uukwaluthi baobab castle was abandoned, its walls crumbling, and the "[Eunda] Headman Shituthi (. . .) informed us he like others made no further forts as (. . .) [the] necessity for such protection had died out".[19] The Ongandjera headman Petrus explained the new dispersed settlement pattern to a 1935 colonial commission: "Before we had the Union [of South Africa] Government here we had a lot of trouble and each man had to try and save his property himself. But now every man is free and can go where he likes".[20]

---

20 July 1948 and appendix, Foreign Natives in Ovamboland: Summary: Ondonga Tribal Area, Chief Kambonde to NCO, Okaroko June 1948; NAO 90, statement Aihuki Tolongele, Ondangwa, 5 May 1948; NAO 91, statement Agustinhu Kapetango [March 1954?]; NAO 92, affidavit Gallation Shidive, Ondonga, 25 Feb. 1954; NAO 91, statements Thomas Kupila, 5 May 1954 and Titus Muatelai Kakonda, Ondangwa, 10 May 1954.

[19] NAN, RCO 8, extract RCO's Personal Diary, 13 March 1917, and RCO to Sec. SWA, 27 Oct. 1918, extracts diary, 9–10 Oct. 1918.

[20] NAN, A450, 12, SWA Commission: Minutes of Evidence, Sitting at Ongandjera, Headman Petrus, p. 700; RCO 4, NCO to Sec. SWA, report on Ipumbu, Ondonga, 6 Jan. 1922. Ondonga was

To chiefs, headmen and by extension the colonial officials who relied on them to administer Ovamboland, the scattering of settlements constituted a challenge for at least two reasons. First, subjects who moved into wilderness areas beyond the old settled core districts effectively moved outside of colonial control. Second, settlers from neighboring districts moved into the middle floodplain wilderness and claimed it as their own. As settlers reduced the wilderness belts and land became scarce, kings and headmen staked their claims to previously unsettled land in what appeared a 'land speculation' frenzy: the more land a king or a headman could amass to allocate, the more followers he could attract (and followers could be taxed and mobilized for labor), and the more grazing areas and water he controlled for his and his followers' livestock.[21]

What the colonial officials of Ovamboland began to consider 'illegal settlement' in the wilderness areas came to be seen as a major problem by both the colonial administration and local kings and headmen. The Ovamboland annual report for 1941 stressed that "[t]he large number of kraals established in the bush areas, during recent years, is very noticeable". In 1946, the Ovambo kings and headmen urged Native Commissioner C.H.L. Hahn to strengthen their authority outside the settled zones proper and Hahn was more than sympathetic:

> The leading natives one and all complain of the growing inclination of their subjects to establish themselves outside the proper tribal area in unauthorised bush country. They are taking steps to have these people moved to where there are fixed settlements and proper tribal control.[22]

In 1947, for example, an Ombalantu headman requested the personal intervention of the Native Commissioner because he was unable to prevent people from cutting down trees to clear new farms and fields.[23]

---

more secure somewhat earlier and the resulting colonization of wilderness areas occurred earlier, see RCO 3, appendices to British Consul, Luederitzbucht, 14 Feb. 1913.

[21] See, for example, NAN, A450, 7, Annual Report Ovamboland 1937, and NAO 99, Headmen Ombalantu to NCO, Ombalantu, 9 Aug. 1951.

[22] NAN, NAO 21, Quarterly Report Ovamboland, April – June 1946.

[23] NAN, NAO 98, Muanyagapo Mbunda to NCO, Ombalantu, 8 April, 14 July and 30 Aug. 1947, and NCO to Muanyagapo Mbunda, Ondangwa, 15 April 1947.

As migrants settled beyond the old population centers during the 1930s, 1940s and early 1950s, 'intertribal' conflict over wilderness territory increased and kings and headmen often called on the Native Commissioners to support their claims. During the 1920s, disagreement about settlement in the wilderness that separated the refugee districts of Oukwanyama and Uukwambi turned violent. The Ondonga and Uukwambi kings demanded that a "definite and visible line [be] demarcated" to prevent any further Oukwanyama expansion southwards into their territories. In 1941, the colonial administration resorted to drastic measures to discourage encroachment on 'wilderness' borderland: it evicted six households that had settled in the 'uninhabited bush' areas of Uukwambi and Onkolonkathi and destroyed the homesteads. By 1948, conflicts between Ombalantu and Uukwaluthi settlers who encroached on disputed 'bush' border territory were frequent.[24] In 1953, the Native Commissioner reported to his superior that "[a]ll the tribal borders in Ovamboland are marked by trees being blazed, by Omurambas [seasonal watercourses] or by sections of roads. Many disputes about these borders have arisen (...). It would be useless and costly to fence these borders".[25]

The colonization frontiers cut into the wilderness even though water was in short supply and the construction of new farms and fields was laborious, slow and difficult.[26] One resource that initially seemed available in unlimited quantities in the wilderness was wood. Even in the Oshigambo Valley of eastern Ondonga, which saw a rapid increase of its population as a result of the influx of refugees, land and wood were still abundant in the early 1920s.[27]

---

[24] NAN, RCO 4, NCO to Sec. SWA, report on Ipumbu, Ondonga, 6 Jan. 1922; Manning's memo re. Ipumbu, 19 Dec. 1921 and annexes 1–3, and Manning and Native Commissioner SWA to Sec. SWA, 1 Aug. 1923, and sketch maps; NAO 10, Olli Suikkanen (for Sheja) to Hahn, Ongandjera, 6 July 1932, Administrator to NCO, Windhoek, 25 April 1931, and ANC to NCO, 31 Oct. 1940; NAO 18–21, Monthly Reports Ovamboland, Nov. 1926, June – July 1927, Feb. 1933, Jan. – Feb. 1941, Quarterly Report Ovamboland, April – June 1946; A450, 7, Annual Report Ovamboland 1941; NAO 98, Dalengelue Aitana to NCO, Ombalantu, 21 Jan. 1948, NCO to headman Dalengelue Aitana, Ondangwa, 26 Jan. 1948, and NAO 99, Council of Headmen to NCO, Ombalantu, 15 and 17 Jan. 1952; NAO 51, Meeting at Ukualuthi, 17 Feb. 1955.

[25] NAN, NAO 59, NCO to CNC, Ondangwa, 7 Aug. 1953.

[26] NAN, A450, 9, Tribal laws and customs of the Ovambos; NAO 91, ANC to NCO, Oshikango, 29 Oct. 1953; NAO 100, Chief Kambonde to NCO, Ondonga, 8 Aug. 1952 and Statement Holongo Amshelelonanda at Ondangwa, 4 July 1952.

[27] Interviews by author: Paulus Nandenga, Oshomukwiyu, 28 April 1993, and Nahandjo Hailonga, Onamahoka, 4 Feb. 1993.

To relieve the congestion in Ovamboland, the colonial administration encouraged the colonization of what became known as 'eastern Ovamboland' east and outside of the floodplain. The settlers faced enormous challenges in eastern Ovamboland, not least because of the lack of water resources and the abundance of wild animals that threatened their lives, their livestock and their crops. Not surprisingly, it proved difficult to recruit volunteer pioneers.[28] The eastern frontier leapfrogged along three parallel axes. The northernmost route was the border road / clearing that marked the Angolan-South West African (Namibian) boundary. The Eenhana road was a second axis of advance. The Ondonga-Kurungkuru road that was cleared during the 1920s formed a third route to penetrate the eastern wilderness.[29]

Tree castles and deforestation in the 1920s to 1940s

The pace and the extent of forest clearing in Ovamboland in the 1920s and 1930s astounded colonial officials and missionaries. When South African officers occupied the territory in 1915, only Ovamboland's southern floodplain had been densely settled, while the middle floodplain and the vast expanses east of the floodplain appeared to be virgin wilderness.[30] The most massive population movements from the Portuguese-occupied northern floodplain into the middle floodplain took place during the 1910s and 1920s. The scale of settlement and the subsequent high demand for construction materials in some areas may have quickly depleted the preferred wood resources, leading to the use of alternatives. The latter was more likely a problem in the southern floodplain, which received a large influx of refugees, especially in Ondonga, where wood resources already had been in more limited supply before 1916. Subsequently, in the 1930s, the

[28] Kreike, *Re-creating Eden*, chap. 7. See also NAO 9, NCO to Sec. SWA, Ondangwa, 6 Dec. 1932 and NAO 19, Monthly Report Ovamboland, Oct. 1934.

[29] NAN, NAO 18, Monthly Reports Ovamboland, June 1926, Dec. 1927, Feb. – March and May 1928; NAO 10, ANC to NCO, 30 July 1940, Report on Development Work Undertaken in Eastern Ukuanyama during 1940, and O/C NAO to NCO, 15 March 1940; KAB 1, Submission to Administrator, Secretary and Attorney-General of SWA, 1927; NAO 105, Diaries NCO, Diary 1928, 16 March and 28 Aug. 1928. See also Kreike, *Re-creating Eden*, chap. 7.

[30] Kreike, *Re-creating Eden*, chap. 2.

South African colonial administration temporarily restricted the further influx of Angolan immigrants, especially into the new colonial Oukwanyama district in the middle floodplain. In the 1940s, the Ondonga king threatened to evict a large number of Angolan 'squatters' from his district because of land shortages.[31]

The refugees and migrants who settled the wilderness areas of Ovamboland continued to construct elaborate fortified homesteads that consumed large amounts of woody vegetation. In 1925, an Anglican missionary wrote:

> The kraals are built of poles standing 8 feet high on end, and closely tied together and supported by horizontal poles. Their diameter varies up to 100 yards, and some even larger. This (...) is (...) of great strength against raiding parties, but very wasteful of wood.[32]

And the District Surgeon for Ovamboland observed in 1937:

> The kraal is a relic of the olden days (...) and (...) was always built with a defensive purpose in view. It is more or less circular and averaging from 20 to 120 yards in diameter according to the status of the kraal head. It is generally surrounded by a stockade of poles about 8 to 9 feet high.[33]

Anglican missionaries commented that "[the] waste of timber in a matter of 25 to 50 years will extend the thirst belt and affect the rainfall and it is the duty of the Government and missionaries to encourage tree planting".[34] A newly

---

[31] NAN, NAO 19, Monthly Report Ovamboland, Sep. 1934; KAB 1, Volkmann, Report on the Agricultural and Political Conditions at the Angola Boundary, 30 Oct. 1928; A450, 7, Annual Report Ovamboland 1940; A450, 10, Draft Annual Report Ovamboland 1942; Kreike, *Re-creating Eden*, chaps. 3, 4 and 6. See also NAO 51, Tribal Affairs Ondonga, Chief Kambonde to NCO, Okaroko June 1948, and Ondangua tribal area, 20 March 1950; NCO to Chief Kambonde, Ondangwa, 29 Aug. 1949; NCO to CNC, Ondangwa, 25 Feb. 1949; [CNC] to Sec. SWANLA Grootfontein, 16 Sep. 1948; Cope to Sec. SWANLA, n.p., 9 Aug. 1948; Sec. SWA to Administrator SWA, 3 Aug. 1948; NCO to CNC, Ondangwa, 20 July 1948 and appendix, Foreign Natives in Ovamboland: summary: Ondonga tribal area; Public Service Inspector to Sec. SWA, Windhoek, 30 June 1948; NCO to Recruiting Officer SWANLA, Ondangwa, 10 July 1948; Recruiting Officer SWANLA to Sec. SWANLA, Ondangwa, 22 June 1948.

[32] NAN, A450, 9, newspaper clipping of an article by Tobias [1925].

[33] NAN, NAO 36, Annual Health Report Ovamboland 1937.

[34] NAN, A450, 9, newspaper clipping of article by Tobias [1925]; NAO 26, Report Ovamboland Cotton Prospects, appendix to Alec Crosby to Bishop of Damaraland, St. Mary's Mission, 11 Jan. 1924; NAO 104, SWA Commission 1935, W.E. Cawthorne, Priest in Charge Holy Cross Mission Ovamboland, 14 Aug. 1935, to the Chairman, The Constitutional Commission, Windhoek.

appointed Assistant Native Commissioner for Oshikango in 1931 expressed shock concerning the extent of deforestation in the Oukwanyama district.[35] But his superior, Native Commissioner Hahn, believed these concerns to be exaggerated and opposed any restrictions on wood use by the inhabitants of Ovamboland because he feared it might trigger political instability.[36]

Hahn thought that it was politically inadvisable at the time to impose *any* real conservation regulations regarding tree use, with the exception of limits on the use of timber trees for constructing new missions and churches.[37] Not only did Hahn oppose the introduction of conservation measures to limit clearing woody vegetation to construct new homesteads, but he also encouraged the continued or renewed construction of fortified homesteads. In 1934, for example, the Native Commissioner urged Ovambo leaders to construct and maintain larger homesteads:

> Natives have been encouraged to build bigger and more substantial kraals. In certain tribal areas it has been found that kraals are becoming smaller and smaller and more dilapidated (...). It is in keeping with native order that the chief and headmen live in big and strongly built kraals (...). It is the big native kraals occupied by wealthy and influential natives which retain tribal order and discipline.[38]

Native Commissioner Hahn blamed Christianity and its crusade against polygamy for a decrease in household size and a commensurate decline in homestead size and quality. A monogamous Christian household, he argued, was simply short of hands and could only maintain a part of the former grand homestead.[39] In brief, Hahn's colonial policy is derived from his understanding of the importance of social relations, with consequences for local decisions about the use of human and woody vegetation resources.

---

[35] NAN, SWAA 3, O/C Oshikango to NCO, Oshikango, 17 March 1931.

[36] See chapter 3.

[37] NAN, SWAA 3, NCO to Sec. SWA, Ondangwa, 20 April 1931; NAO 101, NCO to CNC, Ondangwa, 17 April 1948; BAC 131, Agricultural Officer Ovamboland to Bantu Commissioners Ondangwa and Oshikango, [Ondangwa?] 28 Jan. 1957. See also chapter 3.

[38] NAN, A450, 7, Annual Report Ovamboland 1935.

[39] NAN, A450, 7, Annual Report Ovamboland 1937.

Colonial concerns about overpopulation and deforestation in the 1950s

In the 1950s, a new generation of colonial scientists raised anew the alarm about overpopulation. The medical officer for Ovamboland in 1953 painted the specter of overpopulation, deforestation and deteriorating health conditions:

> The habitable areas are limited chiefly by the terrain and the water supply. Hence the density of the population is already becoming a problem in some areas. Scarcity of durable and suitable building materials for the construction of kraals, and the overgrazing of areas denuded of trees, will amongst other things interfere with the water supply and multiply the danger of fly and tick borne diseases.[40]

In 1956, the administration's newly appointed Agricultural Officer concluded that the two most populous and largest districts of Ovamboland—Oukwanyama and Ondonga—were overpopulated and in the throes of deforestation. Worst off was the area around Ondangwa, which already was denuded of trees. The most densely settled parts of the Uukwambi, Ombalantu, Onkolonkathi and Eunda districts were "populated to their full capacity. Any further increase of kraals, fields, and livestock will result in overpopulation".[41]

Two years earlier, when the highest South African colonial official for Namibia, the Administrator for South West Africa, visited Ovamboland, he admonished the audience at a 'tribal meeting' in Oukwanyama: "You should not cut these beautiful trees. Cut the ugly or dry trees".[42] In 1956, when the colonial administration for the first time addressed all of Ovamboland's chiefs and headmen in a public meeting, the officials advised against clearing new homesteads in wilderness areas because the areas were to become forest reserves: "Chiefs and headmen are instructed not to allot sites in wooded areas, but only in strictly residential areas".[43] Forest

---

40 NAN, NAO 65, Annual Health Report Ovamboland 1953.

41 NAN, BAC 133, Van Niekerk, Travel Report Ovamboland for June 1956 to CNC, Ondangwa, 19 Nov. 1956, and Agricultural Officer to NCO, Report of travel to the northwestern part of Ovamboland from 20–22 June 1956, Ondangwa, 4 July 1956.

42 NAN, NAO 64, Minutes of Ukwanyama Tribal Meeting [12 July 1954].

43 NAN, BAC 44, Minutes of the first general meeting of Chiefs and Headmen of all tribes in Ovamboland, Ondangwa, 22 Aug. 1956; BAC 133, Agricultural Officer to NCO, Report of travel

reserves, where tree felling was prohibited and only dead wood could be gathered for domestic purposes, however, did not materialize until decades later, and they proved ineffective. Proclaiming the remaining wooded wilderness areas as forest reserves floundered in the face of disputed claims over wilderness areas between various chiefs and headmen, which prevented delimitating the borders of the colonial districts.[44]

Ovamboland's colonial chiefs and headmen continued to have an interest in limiting further settlement in the remaining wilderness areas because the activities of people in those areas were more difficult to exploit and control. In this respect, their interests coincided with those of the colonial administration, but collided with the interests of subjects who wanted a farm of their own. For example, during a meeting in Ondonga in 1961, when the Bantu Commissioner warned the headmen that trees and forests should be protected against deforestation, Amtenya Shenuka spoke up from the audience and stated: "regarding new homesteads in the forests. There are many young people without homesteads".[45] The Ondonga king proclaimed at the same meeting that further expansion in western and northern directions would be prohibited to ensure future wood supplies; new homesteads could only be made in the eastern and southern directions where sufficient land was available. An Uukwaluthi headman pointed out that laws existed that prohibited clearing the forest in the border areas between the different colonial districts and complained that "[p]eople move away and they should return to their land in the inhabited area, where spots to make homes are abundant without the need to cut down the forest". The Uukwaluthi king agreed and admonished: "Listen well (. . .). no forests can be eradicated".[46]

Although Ovamboland's kings and headmen supported colonial officials' attempts to conserve forests *outside* the village zones, they were not on the same

---

to the northwestern part of Ovamboland from 20–22 June 1956, Ondangwa, 4 July 1956 and Agricultural Report Ovamboland 1955–1956.

[44] NAN, BAC 133, Van Niekerk, Travel Report Ovamboland for June 1956 to CNC, Ondangwa, 19 Nov. 1956, and Agricultural Officer to NCO, Report of travel to the northwestern part of Ovamboland from 20–22 June 1956, Ondangwa, 4 July 1956.

[45] NAN, BAC 131, Quarterly Report Meeting held at Outanga, Oundonga [Ondonga], 27 Dec. 1961.

[46] NAN, BAC 131, Quarterly Report Meeting held at Outanga, Oundonga [Ondonga], 27 Dec. 1961. The same problem occurred in Ongandjera, BAC 44, Quarterly Meeting Ongandjera, 28 Dec. 1961.

page regarding forest conservation policies *within* the village territories that they controlled. Chiefs and headmen were eager to attract followers by offering farmland in order to expand their income base for labor, taxes and levies (including land fees). They therefore resisted imposing limits on the number of households and farm plots per village.[47] Moreover, wood within the villages could be used to construct and maintain farms, although felling trees for homestead poles formally was limited to "only the useless and, where possible dead trees (...). Each application should be referred to the Headman".[48] In 1961 Uukwambi district, for example, growing trees could not be cut down for poles or other purposes.[49]

Population growth in Ovamboland

How 'real' were the mid-1950s colonial concerns about overpopulation? Population figures supplied by the Ovamboland administration must be used with care. Before the 1960s, the Ovamboland administration conducted an actual population count on only three occasions: in 1933, in 1938 and in 1951. All other figures in Ovamboland's reports until the mid-1950s were based on these three counts. In the annual reports, officials either simply repeated the numbers from the last report, or they added a percentage to the figures each year based on the assumption that the population was growing at a certain rate. For example, in 1946, Native Commissioner Hahn complained about the difficulty of conducting an actual

47 NAN, BAC 44, Minutes of the First General Meeting of Chiefs and Headmen (...), Ondangwa, 22 Aug. 1956; BAC 133, Agricultural Officer to NCO, Report of travel to the northwestern part of Ovamboland from 20–22 June 1956, Ondangwa, 4 July 1956, Van Niekerk, Travel Report Ovamboland for June 1956 to CNC, Ondangwa, 19 Nov. 1956, and Agricultural Report Ovamboland 1955–1956; NAO 64, Minutes of Ukwanyama Tribal Meeting [12 July 1954]; BAC 131, Quarterly Report Meeting held at Outanga, Oundonga [Ondonga], 27 Dec. 1961, and Minutes of Quarterly Meeting in Uukwambi, 18 Sep. 1961; BAC 44, Quarterly Meeting Ongandjera, 28 Dec. 1961; AHE (BAC) 1/2, minutes of a meeting at Okalongo, 11 Jan. 1965; Joshua Mutilifa interview by author, Omhedi, 8 March 1993.

48 NAN, BAC 44, minutes of meetings held at Four Centers in the Ukuanyama Tribal Area, 7–21 June 1957. In 1970, an Ovambo Homeland Administration Committee recommended that residents should continue to be allowed to cut down trees to construct and maintain farms, NAN, OVJ 15, minutes of the elected Committee on Land Ownership and Use, Oshakati, 4 Dec. 1970, appendix to Secretary of the Interior to Secretary Justice and Labor, Ondangwa, 9 Nov. 1973.

49 NAN, BAC 131, Minutes of Quarterly Meeting in Uukwambi, 18 Sep. 1961.

census in Ovamboland and eventually submitted estimates rather than actual counts on the required census sheets. In 1948, Hahn's successor Eedes simply copied the population figures in his predecessor's last annual report.[50]

According to the colonial and later figures, the population nearly doubled from 107,861 to 200,253 people between 1933 and 1951, and subsequently tripled to 618,669 between 1951 and 1991.[51] Interestingly, the figures suggest a relative decline in population growth after 1951: had the population been calculated to have increased at the same rate between 1951 and 1991 (a forty-year period) as it had in the less than two decades leading up to 1951, the figure in 1991 would have been 800,000 people.

Although fertility figures (the number of live-born children per woman) are rare, data for the Oukwanyama district for the 1930s provide some clues to understanding the area's population dynamics. The 1938 census recorded 4,600 infants (defined in the census as children of two years of age and under), suggesting that on average approximately 2,300 children were born in 1936 and in 1937. The population of Oukwanyama increased by 11,000 persons between 1933 and 1938, and 2,300 births per year would account for the increase, although these figures do not reflect mortality. Infant mortality was high, and actual natural increase was much lower than the above figures at first glance seem to suggest. The 1934 figures for Oukwanyama, for example, recorded 663 births, 245 infant deaths and 295 other deaths. Natural increase of the population thus was only 123 people, i.e., three per thousand based on the population figures for 1933.[52] In 1933–1934, Ovamboland was still in the aftermath of a severe drought, which must have depressed the number of births and increased mortality, so the figure may be on the low side. It nevertheless suggests that the population increase of 11,000 people in Oukwanyama district between 1933 and 1938 principally was due to immigration rather than to natural increase. Figures for Ombalantu, Ongandjera and

[50] See NAO 24, 1933 Census and 1938 Census; NAO 19, NCO to Sec. SWA, Ondangwa, 15 Feb. 1934; NAO 20, Annual Report Ovamboland 1938; NAO 23, NCO to Sec. [SWA], Population Census 1946, Windhoek, 8 March 1946, and Final Return Census 1946, Province of South West Africa, Ovamboland, 15 May 1946; NAO 61, Annual Reports Ovamboland 1948, 1950 and 1951.

[51] NAN, NAO 24, 1933 Census and 1938 Census; NAO 61, Annual Report Ovamboland 1951; Republic of Namibia, *1991 Census*.

[52] NAN, NAO 23, O/C Oshikango to NCO, 30 March 1938, and NAO 24, O/C Oshikango to NCO, Oshikango, 14 May 1935.

Onkolonkathi underscore that in-migration may have been an important source of population growth between 1933 and 1938. No comparable figures are available for Ondonga, but even if Oukwanyama was the only colonial district where migration determined population dynamics, its overall impact was critical: in 1933, the district accounted for nearly 40 % of the total population of Ovamboland.[53]

Figures from the 1991 census suggest that natural population increase was significant in the 1940s and 1950s. Only 8 % of the survivors born between 1947 and 1976 were foreign born, a percentage that is lower than that for the older age groups. "Foreign born" refers to refugees from the northern floodplain in modern Angola.[54] Ovamboland had a young population in 1991: over one-third of the population had been born after 1976 and the survivors from the 15–24 age category were twice as numerous as survivors from the 25–34 age category.[55] The data from the 1991 Namibian census also indicate that the fertility of women born between 1932 and 1946 was notably higher than that of the older age groups. In turn, the average number of live-born children for women born between 1927 and 1931—although lower than the 1932–1946 age group—was substantially higher than that of the 65+ group that had been born in 1926 or earlier. The women with the higher number of live-born children are likely to have begun to give birth in the late 1940s and early 1950s, and they subsequently would have produced a baby boom until well into the 1960s.[56]

Malthusian population bomb mechanics caused by a natural population growth only became a factor after World War II. Prewar population increase in north-central Namibia was the result of massive in-migration of refugees from the northern floodplain in modern Angola. The resulting population pressure on woody and other environmental resources in north-central Namibia was thus not

---

[53] For the census figures, see NAO 24, Ovamboland Census 1933 and 1938 Census. For the 1936 data on births and mortality, see NAO 23, Mateus Angolo [Ongandjera] to Songola [Hahn], 28 Dec 1936, and Festus Hango [to Hahn], n.d.

[54] Republic of Namibia, *1991 Census*, Report A, Statistical Tables, vol. 1, table A03.

[55] Ibid.

[56] Ibid., Report A, Statistical Tables, vol. 4, table F01. The increase in the averages of the total number of children born alive per age group as derived from the table is clearly discernible in the figures for Enumeration Areas Oshakati and Ondangwa (that cover Ovamboland). The average for the 65+ group in Oshakati, 6.17, increased to 6.4 for the 60–64 group (born between 1927 and 1931) and rose to 6.7 for the 50–59 and 50–54 groups, and to 6.53 for the 45–49 group. The average percentage for the Ondangwa area for the 65+ group was 5.98, 6.5 for the 60–64 age group and approximately 7 for the age groups 55–59, 50–54 and 45–49.

the result of natural causes, but was a consequence of political insecurity in the Portuguese colony of Angola. It was also not a mechanical and linear process; rather the decision to flee to the Namibian side or remain on the Angolan side was made by individuals and households. Instead of an explosive population growth it was the displacement and reconcentration of the existing population of the floodplain within the larger region that transformed how, where and to what extent woody vegetation resources in north-central Namibia were utilized.

Woody vegetation resources by the close of the twentieth century

Poles and firewood were scarce in Ombalantu district by the early 1970s. By the early 1990s, the same applied to much of the central floodplain in the peri-urban area around Ondangwa and Oshakati. Households in villages close to the border had access to abundant wood resources in Angola's northern flood-plain. In villages further south and closer to Ondangwa and Oshakati, including, for example, Oshomukwiyu, Omupanda and Eko, however, firewood was in such short supply that people dug out and used old tree stumps and tree roots. In Oshomukwiyu, the only nonfruit trees left in the landscape consisted of sparse, heavily coppiced mopane stumps and some mopane bush.[57] While a bundle of firewood from Angola fetched two rand in border villages in 1993, it fetched five rand in the Namibian border town of Oshikango; a bundle of fire-wood was sufficient to meet a household's cooking fuel requirements for two or three days.[58] By the early 1990s, a variety of woody species, including mopane (*Colophospermum mopane*), red bushwillow (*Combretum apiculatum*) and wild seringa (*Burkea africana*) were used as firewood.[59] Southwest of Oshakati, one of

---

[57] NAN, OVA 57, Dr. H.A. Lueckhoff, report on a visit to South West Africa, Nov. 3–15, 1969, appendix Regional Forester to Director-in-Chief Department of Bantu Administration and Development Pretoria, Grootfontein, 3 April 1970 and interviews by author: Paulus Nandenga, Oshomukwiyu, 28 April 1993; Kulaumoni Haifeke, Oshomukwiyu, 11 May 1993; Lea Paulus, Onandjaba, 17 June 1993; Johannes Abraham, Odibo, 20 May 1993; Personal observations by the author, Oshomukwiyu, 27 April and Eko, 25 May 1993.

[58] Johannes Abraham, informal interview by author, Odibo, 20 May 1993.

[59] Johannes Abraham, informal interview by author, Odibo, 20 May 1993.

Ovamboland's two largest towns, only fruit trees remained and people used palm fronds and dried dung as fuel. A 2001 study, however, claimed that wood use in Ovamboland as a whole was sustainable, irrespective of the population growth.[60] Despite an undeniable shortage of wood resources in parts of Ovamboland, the dire predictions that Ovamboland would degenerate into a desert had not materialized by the close of the twentieth century. Why not? First, such predictions often were overstated. For example, a warning by the forester Dr. Lueckhoff that Ovamboland was transforming into an inhospitable desert was based on a tour through Ovamboland in late 1969—during the height of the dry season! He pointed to 'treeless plains' northeast of Oponono Lake as evidence and considered the area's sparse trees as relic vegetation of a previously more abundant tree cover.[61] Yet, earlier descriptions of the area depict it as grass plains with little woody vegetation. Colonial officials also presumed that 'Africans' had a negative attitude towards trees and they explicitly attempted to re-educate African subjects on the value of trees, for example, in 1972, when the administration embraced reforestation under the South African 'Our Green Heritage' environmental awareness campaign.[62]

Second, woody vegetation became less exclusively a source of construction materials (and protection). The fate of the baobab may in fact suggest a trend for other woody species that served 'protective' functions. Although the baobab castle represented the starkest example of the critical safety functions of woody vegetation early in the twentieth century, by the end of the century its use as a stronghold was a distant memory. Overall, the importance of woody vegetation as an almost exclusive source for construction materials had been declining since the 1950s, when wood began to be replaced by alternative materials. Of a sample

---

[60] Personal communication Joseph Hailwa, District Forester, 24 March 1992, and Namibian Institute for Social and Economic Research, "Namibian Energy Assessment: Household Energy Consumption, Distribution and Supply Survey of the Owambo Region of Northern Namibia and Katatura, Windhoek" (University of Namibia, 1992). Erkkilä concluded that woody biomass consumption for Ovamboland as a whole was sustainable, see Erkkilä, "Living on the Land", p. 100, table 12.

[61] NAN, OVA 57, Dr. H.A. Lueckhoff, report on a visit to South West Africa, 3–15 Nov. 1969, appendix Regional Forester to Director-in-Chief Department of Bantu Administration and Development Pretoria, Grootfontein, 3 April 1970.

[62] NAN, OVA 57, Director of Agriculture Ovamboland to Secretary Bantu Administration Pretoria, [Ondangwa], 11 Oct. 1972, and Secretary Bantu Administration to Director of Agriculture Ovamboland Government, Pretoria, 6 Nov. 1972.

of surveyed households that retained a palisade in 1993, in 13 % of the cases (50 out of 313) the materials were of nonwood origin.[63] Almost 10 % of the households used millet stalks and 7 % used palm fronds, wire or bricks.[64]

Clay bricks and bricks made from a mixture of clay and cement became increasingly common in even the most remote rural areas of north-central Namibia. In 1966, of the 49 homesteads that were razed to make room for Ogongo Agricultural College, 10 % contained brick buildings.[65] In 1967, 231 households received compensation for losses in connection with widening the Ruacana-Ondangwa and Oshivelo-Ondangwa-Oshikango roads. The homesteads of almost half (110) of the households contained one or more brick constructions and 24 (10 %) used corrugated iron as a construction material, mainly for the roofs.[66] In 1993, wood was still a critical material, and two out of every three Ovamboland Multi-Purpose Investigation for Tree-Use Improvement (OMITI) survey households had at least one hut made with a wall of wooden poles. But two out of every three households also had at least one additional hut made with brick walls. The walls consisted predominantly of mud bricks; only one out of every ten households had one or more cement brick huts. Wood- and mud-walled huts were mentioned by one of every three respondents in the 1993 OMITI survey, and corrugated iron by one of every six.[67] Indeed, late in the 1993 rainy season, in the villages of Eko and Omupanda, where construction wood was in very short supply, young boys could be observed making bricks during the school holidays, typically using earth taken from termite mounds, but also using cement.[68] The shortage of construction material was especially obvious in the Ondangwa-Oshakati area, which was sparsely forested earlier in the twentieth century. In 1993, millet and sorghum stalks were

[63] OMITI 4.3.1.

[64] OMITI 4.3.1.

[65] NAN, AHE (BAC) 1 / 346, Bantu Affairs Commissioner to Chief Bantu Commissioner, Ondangwa, 30 Dec. 1965, and Chief Bantu Commissioner SWA to Secretary Bantu Administration and Development, Windhoek, 11 Jan. 1966. On bricks, see also WAT ww17, S. Davis, Tour of Northern Territories—Some Random Comments and Thoughts, and Kaulikalelwa Oshitina Muhonghwo, interview by author, Ondaanya, 2 Feb. 1993.

[66] NAN, OVA 53, Sec. SWA to Sec. Agriculture Owambo, Windhoek, 24 June 1974, appendices A–C.

[67] OMITI 4.3.11. The 1991 census underrepresented the use of nonwood construction materials for huts. The census identifies 598 homesteads in the category "Kraal / Hut" with cement block constructions but it has no category for clay brick constructs. See Republic of Namibia, *1991 Census*, Report A, Statistical Tables, vol. 5, table H04.

[68] Author's personal observations, Eko, 25 May 1993.

used as construction material for huts (mentioned by 4 % of the OMITI sample) and for palisades.[69] Although wood was in short supply in the central areas in the late twentieth century, therefore, environmental change in north-central Namibia cannot be reduced to a unilinear, progressive and irreversible process of deforestation based on the hypothesis of population explosion.

The role of population movements and their impact on the making and unmaking of human-settled areas and uninhabited 'forest,' 'bush' or 'wilderness' areas was critical. Colonial violence and the demarcation of colonial boundaries led to massive flight from the Portuguese-occupied northern floodplain into the South African-occupied middle floodplain, an area that was overwhelmingly wilderness. The subsequent settlement of the middle floodplain wilderness by the refugees from the northern floodplain, and the settlement of the wilderness zones that had separated the precolonial polities from one another by refugees from the northern floodplain and migrants from the old southern floodplain heartlands, dramatically changed Ovamboland's environment.

The impact of population density on the forest environment therefore is ambiguous and beyond being a quantitative and biological factor. Until the 1940s, the 'population' factor in Ovamboland exerted its most important influence through migrations and flight, and not through the mechanics of any 'population bomb'. Thus, security and insecurity concerns contributed critically to *how much* woody vegetation was consumed, and *why*. That population as a qualitative factor is as important as population as a quantitative factor also suggests that population may play a critical role in environmental change even under conditions where overall population-to-land ratios appear to be low.

---

[69] On the use of millet stalks, see OMITI 4.3.11. Millet stalks are not included as a category of building materials in the 1991 census. See Republic of Namibia, *1991 Census*, Report A, Statistical Tables, vol. 5, table H04.

3

Conquest of Nature: Imperial political ecologies

The ideas, policies and practices regarding the conquest, occupation and administration of colonial empires are major factors in understanding environmental change. Perceptions of non-Western environments which involve a set of issues that Grove labeled 'green imperialism' can be distinguished from Crosby's 'biological imperialism' because Grove highlights human agency (i.e., Culture) and ideas while Crosby emphasizes biological agency with humans as the unintentional vectors (i.e., Nature's agency).[1]

Colonial conservation and development priorities and projects shaped the non-Western environment physically and conceptually—often in very dramatic ways. Hunting and gathering forest products, for example, were redefined as poaching when colonial administrators created game and forest reserves.[2] Moreover, as demonstrated in the previous chapter, the insecurity that marked colonial conquest and the draconian punishment meted out to maintain colonial law and

[1] Crosby, *The Columbian Exchange* and *Ecological Imperialism*; Grove, *Green Imperialism*.

[2] See, for example, Grove, *Green Imperialism*; Anderson and Grove, *Conservation in Africa*; MacKenzie, *Imperialism and the Natural World* and *The Empire of Nature*. See also Guha, *The Unquiet Woods*, and Peluso, *Rich Forests*.

order caused population redistributions through flight or migration, which in turn had dramatic environmental consequences. In north-central Namibia, colonial officials increasingly enforced external and internal colonial borders to limit the movement of subjects, animals and goods between different territories and within territories. A policy that had an even greater impact than proclaiming game reserves in the area, however, was the move to limit the mobility of cattle into and outside of Ovamboland in the name of disease control.[3]

A political ecology focus builds on the premise that actions relating to the environment ultimately are dominated or inspired by political and power considerations. Alternatively, political ecology highlights power as a means to attain an environmental objective. In practice, political ecology often highlights politics and power struggles and relegates environmental dynamics to the background. The green imperialism theory emphasizes that environmentalism was created overseas, in the process of empire building, and in interaction with and often dependency on non-Western environmental ideas and practices, rather than being an exclusive product created in the halls of power, offices or labs in the West.

The political ecology framework not only sheds light on power struggles between the colonizer and the colonized, but also on the policies and measures that in turn resulted from power struggles within the 'colonial' and 'colonized' categories. Officials at different levels and in different departments, administrators, scientists, missionaries, settlers, local headmen, Christians of various denominations and non-Christians were often at odds, sometimes motivated by personal rivalries between individuals. This chapter focuses on power struggles that either indirectly affected the environment or were fought over particular environmental resources.

The idea of a post-World War II second colonial conquest of Africa is useful here.[4] The first colonial conquest was the turn of the nineteenth century military and political conquest that established control over the peoples of empire in order to harness their labor. But effective exploitation and development required harnessing Nature's resources as well, especially when officials became concerned about rapid population growth and a limited natural resource base. Diseases and

---

3 See chapter 6.

4 The term "second colonial conquest" was coined by Lonsdale, see Lonsdale, "East Africa", p. 540.

droughts were a further limitation on the efficient use of natural resources. The second, post-World War II conquest of Africa (and Asia) was in effect a scientific conquest of Nature (i.e., Nature as the physical environment). After the first conquest the colonial rulers established new borders affecting environmental use and management (intercolonial borders, Native Reserves, conservation areas) and introduced conservation measures to limit hunting immediately upon conquest. In many areas, the violence of the initial conquest displaced populations and livestock was lost to the conquerors as spoils of war. Because Africans were regarded as part of Nature, conquering African peoples would provide colonial administrations with the means to subjugate Wild Africa.

But in general, before World War II the main colonial policy that shaped environmental dynamics was indirect, that is, territorial control was exercised by instituting political and conservation boundaries and disease cordons, and by confining people and animals to reserves. In pre-1940s Ovamboland, the colonial administration also sometimes directly and purposely interfered in how its inhabitants used the environment, although with limited impact. Conservation measures prohibited hunting in the newly established game reserves (resulting in the Etosha Park) and proscribed the hunting of 'royal game' such as elephants and lions. Overall, in pre-World War II Ovamboland, colonial governance not only was marked by indirect political rule but also by an indirect colonial stewardship of Ovamboland's Nature.

Indirect environmental rule frustrated colonial officials because Africans persisted in what officials regarded as their inefficient and wasteful use of Nature's bounty. After World War II, colonial states directly tackled the African environment, causing a second round of struggles that had as its main objective to conquer Africa's Nature, harnessing its wild resources to subsidize colonial rule and economic development.

In Ovamboland, the two conquests left clear imprints on the landscape. Military and political conquest was enormously destructive. The high levels of sustained violence in the northern floodplain left entire areas depopulated, while thousands fled into the middle floodplain wilderness where within a generation, refugees created a new humanized environment, Ovamboland's Oukwanyama district. The history of the delimitation of the border between the Portuguese colony of Angola (including the northern floodplain and the old Oukwanyama) and the South

African colony Namibia (then known as South West Africa and legally a mandate from the League of Nations) demonstrates the extent to which grand political strategy made in the imperial capitals was written onto the land. The border was disputed and moved over time, leading to further displacements, and causing the creation of yet new village environments deeper in the wilderness.

Science offered both a tool and legitimization to conquer Nature in Ovamboland directly and to domesticate its environment. Especially during World War II, the state demonstrated its strength in carrying out planned and top-down resource allocation. Extending political control over Ovamboland's environmental resources led not only to struggles between colonizers and colonized, but also to competition within the ranks of both groups. Rifts and alliances were fluid. To appreciate the full complexity of how power struggles shaped the environmental dynamics and the other way around, it is necessary to differentiate the processes of change. Various struggles acted upon one another at different times but without being fully integrated, leading to multiple, fleeting outcomes as the relationships and roles of subjects and objects of the power struggles shifted. When Harold Eedes took over as the Native Commissioner for Ovamboland in 1947, he spearheaded the second colonial conquest: the conquest of Ovamboland's Nature. His predecessor C.H.L. Hahn had earned his laurels in the first, military and political conquest of Ovamboland, which, unlike the rest of Namibia, had never been occupied by the Germans. Eedes, seeking to distinguish himself from his personal rival Hahn, presented himself as the consummate modernizer by deploying science, scientific knowledge and scientists to conquer Ovamboland's natural (cattle) disease environment. In the 1920s, Eedes had been Hahn's second-in-command and the Assistant Native Commissioner for Ovamboland stationed at the border with Angola in Oshikango, Oukwanyama district. The two clashed openly in 1929 and 1930 about the severity of the famine conditions in Oukwanyama. Eedes argued that Oukwanyama required immediate famine aid. Hahn had Eedes removed from Ovamboland as a result of this dispute, but to Hahn's chagrin, Eedes was promoted to the rank of Native Commissioner of the smaller Okavango Native Territory east of Ovamboland.[5]

---

[5] NAN, AGR 25, Senior Veterinary Surgeon to Sec. SWA., Windhoek, Nov. 13, 1941.

Eedes was also the last of a generation of colonial officials whose qualifications were based on personal and empirical inductive knowledge—i.e., based on his experience with the Ovambo (like Hahn)—whereas post-World War II experts claimed expertise derived from a scientific deductive knowledge base. Yet until his retirement in 1953, Eedes quite successfully harnessed the scientifically trained veterinary and agricultural experts that were seconded to his staff to serve his own agenda.

The political ecology of insecurity

In the late 1800s and early 1900s, the Ovambo floodplain and the surrounding areas were subject to raiding and warfare linked to the slave trade and imperial expansion, and simultaneously to being plagued by periodic droughts and pestilence. Insecurity in the Ovambo floodplain, as elsewhere in the continent-wide era of troubles, caused populations to concentrate in fortified defensible sites under the protection of strong military leaders. Ovamboland's elite used ivory and cattle to acquire guns and horses that provided an effective means of defense against raids, by, for example, the gun-wielding and horse-mounted Nama from central Namibia and the Portuguese from the right bank of the Kunene River. The guns and horses were not only for defense, but also provided the means to raid others, which led to retaliatory raids, further escalating the violence. As a result, in the course of the 1800s through 1914, not only outlying farms, fields and villages but also entire districts and an entire kingdom were abandoned and transformed into wilderness. In addition, from the 1880s to the mid 1910s, the decline in animal populations that resulted from commercial hunting and the rinderpest epizootic spurred environmental change. Ivory hunting led to a sharp decline in elephant numbers and the rinderpest killed wild and domestic browsers and grazers. As a result, bush vegetation may have outstripped the capacity of browsing animals to check it, resulting in bush encroachment as undesirable bush species invaded park- and grasslands.[6]

---

[6] See Kreike, *Re-creating Eden*, chaps. 1–4.

At the same time, the combined threats of Portuguese and German conquest created an ever-increasing demand for firearms, horses and other war materials. Ovambo elites not only raided neighboring polities and communities but also raided their own subjects, causing further internal insecurity and population displacement. After a series of disastrous defeats, in the first decade of the twentieth century, the Portuguese occupied most of the northern floodplain, with the exception of the powerful Oukwanyama kingdom. When a military force from German South West Africa attacked a Portuguese border post in the northern Ovambo floodplain in 1914, a massive revolt routed the Portuguese. A large expeditionary army sent from Portugal in 1915 reoccupied the lost territories and conquered Oukwanyama, leaving the entire northern floodplain in Portuguese hands. The Portuguese success caused a massive population flight into the middle and southern floodplain, where South African forces had arrived after defeating the German forces in South West Africa / Namibia.[7]

King Mandume of Oukwanyama and many other northern floodplain leaders and a large number of their followers fled to the South African-controlled side, set off from the Portuguese-occupied northern floodplain by a seven-mile band of disputed territory that until 1927 was co-administered by the South African and Portuguese colonial administrations. King Mandume continued to challenge the Portuguese and was defeated in 1917 at Oihole, where he met his death. Although the Oihole battle was the last major armed encounter, neither colonial administration managed to gain full military control of the floodplain until the early 1930s, when King Iipumbu of Uukwambi was bombed into submission and the southern Ovambo floodplain communities were disarmed.[8]

Indirect environmental rule

In 1920, C.H.L. Hahn, a military officer who had participated in the 1916–1917 campaign against King Mandume, became the Native Commissioner of the South African-controlled parts of the Ovambo floodplain, a position he retained

[7] Ibid., pp. 35–55.

[8] Ibid., pp. 57–80. For details about the refugee movements, see chapter 2. On the disarmament, see chapter 5.

until his retirement in 1947. During his tenure, he very much ruled the 'Native Territory' of Ovamboland as his personal fiefdom, submitting regular reports only after he was forced to do so following an investigation into alleged abuses in 1924. Hahn and his successor H.L.P. Eedes (who ruled Ovamboland from 1947 until 1953) considered themselves the paramount chiefs of the territory, which they ruled with a combination of (the threat of) violence and patronage. Hahn prided himself on ruling through the traditional chiefs and headmen of Ovamboland. In 1924, in his first annual report, he explained: "The policy since our establishment has been to allow natives to rule themselves according to ordinary native law, and no attempt has been made to change this or in any way to interfere with native custom".[9] In his 1938 Annual Report, he stated:

> The policy of the Administration in Ovamboland is that of indirect rule in a modified form. It is not nearly so far advanced as a similar policy exercised by the British Government in its East African possessions, but the principal of control is identical, namely to build on the institutions of the people themselves, tribal institutions which have been handed down to them through the centuries (...). The great essential in carrying out a policy based on the foregoing is the education of the native Chiefs and Headmen so that they may administer justly and efficiently under our guidance and supervision.[10]

The latter, he emphasized, required a persistent "personal touch", which he complained he had not been able to focus on optimally due to a shortage of staff and transport. The "personal touch", however, was a euphemism for unadulterated personal rule, and what Mahmood Mamdani has called 'decentralized despotism', a highly authoritarian system of rule.[11]

Hahn did not simply rule through Ovambo traditional chiefs and headmen and traditional law. He usurped what he believed to be the absolute powers of the Ovambo kings / chiefs over their subjects, their material possessions and the environmental resources they relied on, including land. Hahn did not hesitate to use violence or the threat of violence when necessary, resorting to what was then the ultimate new destructive weapon: bombers. Kings Martin of Ondonga and

---

[9] NAN, NAO 18, Annual Report Ovamboland 1924.
[10] NAN, NAO 20, Annual Report Ovamboland 1938.
[11] Mamdani, *Citizen and Subject*, pp. 35–179.

Iipumbu of Uukwambi were too independent-minded for Hahn's liking, especially since their followers were very well armed. King Iipumbu also was not afraid to use his armed retainers to enforce his policies: in 1921, his men destroyed a homestead in the Oukwanyama district that the king felt was intruding on his territory.[12]

In 1925, to intimidate Kings Iipumbu and Martin, Hahn arranged for a visit of two South African military planes to demonstrate the power of aerial bombing which in 1922 had been successfully used to suppress the Bondelswart rebellion in colonial Namibia. One of the two planes sent to Ovamboland was piloted by Colonel Sir Pierre van Ryneveld, an icon in the history of South African military aviation. Hahn arrived in Ondangwa as a passenger on one of the planes and on April 17, King Martin of Ondonga met the pilots and was given a demonstration of the firepower of one of the planes' Lewis machine guns. The next day the planes demonstrated a bombing run, dropping ten bombs. King Martin did not attend the event, but he was sure to have felt the impact of the explosions which reverberated beyond Ondonga district through Uukwambi and Oukwanyama districts. King Iipumbu of Uukwambi mobilized his warriors when he heard about the arrival of the planes; in the evening of April 18, the two war planes circled around King Iipumbu's homestead and airdropped a message from the Administrator of South West Africa (colonial Namibia), which Assistant Native Commissioner Eedes read to the king.[13] In Hahn's perception, the demonstration was successful. He noted in his 1925 Annual Report: "The visit of the aeroplanes had a very marked effect on the natives and the matter is still discussed by them".[14] The effect on King Iipumbu, however, was not permanent. In 1932, Hahn called in warplanes and armored cars, exploiting the available firepower to the fullest. Not only was King Iipumbu exiled and his followers disarmed following the bombing of his palace, but Hahn also used the occasion to persuade King Martin of Ondonga and the leaders of the other southern Ovambo districts to promise to hand in their modern firearms. Hahn thereafter kept the Ondonga airfield well maintained and also established airfields at Engela (near the border)

---

[12] NAN, A450, 7, RCO to Sec. SWA, Ondangwa, 15 Nov. 1921.

[13] NAN, NAO 18, Monthly Report Ovamboland, April 1925. On the Bondelswart rebellion, see Emmett, *Popular Resistance*, pp. 122–124.

[14] NAN, NAO 18, Annual Report Ovamboland 1925.

and Omboloka, in the far east of Ovamboland.[15] In December 1939, Hahn once more resorted to warplanes, this time to impose his will on King Martin of Ondonga, who had refused to relinquish jurisdiction over capital cases to the colonial administration. The arrival of three planes and a strong police unit proved sufficient to allow Hahn to fully disarm the king and his subjects, thus completing the disarmament of Ovamboland almost a quarter century after it first had been occupied by South Africa.[16]

During the 1930s, in the name of defending traditional order and law in Ovamboland, Hahn also crusaded to limit the influence of the missions and its adherents, notably the Finnish Mission Society (FMS) and its new elite of Ovambo pastors and teachers, who openly rejected the authority of the traditional leaders and rules. Hahn had warned that as a result "from an administrative point of view the control over the natives, which at present is exercised through their own chiefs or other ruling natives, will in future become more and more difficult, complex and costly".[17] In 1935, Hahn's superior, the Administrator of South West Africa, advised his subordinate about how to prepare for the upcoming visit of a commission of inquiry, the South West Africa Commission, and what to stress in his testimony:

> You should dilate on the point of the economic completeness of the Ovambo people and the dangers of breaking up their tribal system through a wrong system of education and misapplied christianity. Between ourselves, Judge van Zyl [a member of the commission] asked me what we had accomplished in the way of education after 21 years, and why we had left this matter entirely in the hands of German missionaries. I told him this was a very pertinent question from the political point of view but I should prefer to answer the question after he had been round the country when he would be in a better position to appreciate our arguments. For instance I told him that we might argue that as far as Ovamboland went education as at present understood by our Union [of South Africa] Education authorities was unnecessary and in fact dangerous. What we had done in Ovamboland so far was to enforce peace, law and order which the native population had never known up till now and to assist them to improve their country so as to carry a larger population which naturally

---

15 NAN, NAO 25, NCO to CNC, Ondangwa, 16 Feb. 1943; NAO 20, Annual Report Ovamboland 1937 and Monthly Report Ovamboland, Aug. 1937.

16 NAN, NAO 20, Monthly Report Ovamboland, Dec. 1939. See also Emmett, *Popular Resistance*, pp. 203–205.

17 NAN, NAO 18, Annual Report Ovamboland 1927.

> followed with the cessation of tribal wars and the influx from Portuguese territories (...) as the result of improved conditions here. We had even gone so far as to transport to Windhoek an Ovambo kraal to show the Europeans and natives here how the Ovambos could maintain themselves without having to rely on European support in a country far less fertile than the Territory proper. The christianising of Ovambos, if it resulted in their detribalization would probably result in a few years in our having to feed the population (...). Curiously enough Dr. Holloway [another member of the commission] said the Economic Commission on which he served in the Union had found that a great deal of harm had been done in the Union by misapplied education. I told him he would find interesting scope for study in Ovamboland (...). So you see you have ample scope for your pet views.[18]

The Administrator also supplied Hahn with a list of people who should be asked to testify.[19] Encouraged by this advice, Hahn went before the commission to reject the missionaries' credentials as a source of knowledge about Ovambo society and culture and stressed that he was the sole uncontested authority because of his deep knowledge of the local culture and its traditions. Hahn stated: "they [the missionaries] do not go about. I do not think that there is any Missionary here today who knows the country better than I do". Responding to the concerns of the commission about the contradictions between his evidence and that of the missionaries, Hahn attributed it to the latter's highly localized knowledge. He argued that their experience with Ovamboland's society was limited to the main mission stations and a few outstations and that they had a road bias: "They just go along the main road and they travel from the mission station to the store or the hostel".[20]

Moreover, in his testimony, Hahn warned that the missions—especially the FMS, which was the largest—undermined traditional authority in Ovamboland through their "native teachers" who "catch hold of the children and hold them", and undermined the "tribal order" because they no longer listened to their elders. Experimenting with giving the inhabitants of Ovamboland more administrative powers, he continued, was also dangerous because of the situation just across the border in the Angolan Lower Kunene "where after all you have the same native.

---

18 NAN, NAO 104, Administrator SWA to NCO, Windhoek, 24 July 1935.

19 NAN, NAO 104, Administrator SWA to NCO, Windhoek, 24 July 1935.

20 NAN, A450, 12, SWA Commission: Minutes of Evidence (1935), vol. 12, session at Ukualuthi, 13 Aug. 1935, pp. 659–660.

He is being detribalised at present. The Portuguese Government does not believe in allowing the natives to go ahead according to their own customs. They have a number of headmen who control areas only in so far as the tax collection goes. Further they have no right to try cases or disputes".[21]

With the blessing of the Administrator and having impressed the commission of inquiry, Hahn could implement his pet views. Employing bullying and blackmail, by the end of the decade Hahn had forced the FMS Christians into submitting to the authority of the "traditional" leaders, his chosen headmen. Hahn's notes of his meeting with the leading FMS missionary Alho about the payment of cattle as bridewealth are illustrative:

> Alho accepts the enforcement of custom as long as it is not a bargain!! i.e. Father-in-Law should not pick & chose the ox brought by prospective son-in-law, but should accept the ox as an ox tendered as a token of impending uniting of two groups. He adds that a calfox or old and decrepit animal would not be acceptable (...). I pointed out that the ox has never been tendered as payment since no Ovambo would sell his child for one ox and a few hoes (...). After an argument lasting for about 3 hrs Mr. Alho submitted to my suggestion & 'threat' that I would get the Adm. [Administrator] to support the headmen.[22]

An additional threat Hahn used was that the headmen—who allocated plots of land to newly married couples—would withhold land from recently married Christian couples.[23]

In addition to eliminating challenges to the authority of the headmen, Hahn promoted his own favorites to key positions. He personally chose the members of the new Councils of (Senior) Headmen that he introduced, first in the Oukwanyama district, and subsequently in all of the districts. He placed the Senior Headmen (also called Councilor Headmen or Tribal Councilors) over the remainder of the headmen in matters of administration, justice and the allocation and management of land and other environmental resources in each district. After the violent disposal of King Iipumbu of Uukwambi, Hahn reorganized the district,

---

[21] NAN, A450, 12, SWA Commission: Minutes of Evidence (1935), vol. 12, session at Ukualuthi, 13 Aug. 1935, pp. 666–667.

[22] NAN, NAO 9, notes of a meeting of Hahn with FMS missionaries regarding position of V. Alho, 17 May 1938, Attached to O/C NAO to NCO, Oshikango, 22 Feb. 1938.

[23] NAN, NAO 20, Annual Report Ovamboland 1938; Kreike, *Re-creating Eden*, pp. 117–121.

elevating fourteen of its headmen to Councilor Headmen and allocating between six and nineteen of the remaining headmen as "subheadmen" to each of his chosen Councilors.[24]

In the Oukwanyama district, where his subordinate Eedes pioneered the institution of the Tribal Council, Hahn initially had favored mainly former opponents of the last Oukwanyama King Mandume. But in the mid-1940s and early 1950s, the most powerful Senior Headman of Oukwanyama (the most populous of all the districts of Ovamboland) was Nehemia Shovaleka, who began his career in the 1920s as one of Eedes' 'police boys'. The administration's 'boys' served as messengers, interpreters and, most importantly, as thuggish enforcers of colonial rule.[25] Hahn ruthlessly neutralized traditional leaders whom he considered to be a threat, including Kings Iipumbu and Martin. In 1944, when a man turned up claiming to be Nanjungu Sapetama, the uncle of the late King Mandume, Hahn allowed the man to live with Nanjungu Sapetama's mother, Nawanga. But he warned the seventy-year-old Nanjungu "that any move on his part which is contrary to the established authority will result in his immediate expulsion from Ovamboland and finally that it be made generally known that he is an imposter and not of royal descent".[26] The fact that Nawanga took the old man in as her own son suggests that he was probably not an imposter, but, as the maternal uncle of the former King Mandume, Nanjungu Sapetama would have been the legitimate successor to the last king of Oukwanyama, throwing a wrench into Hahn's system of personal and indirect rule through the Tribal Councilor headmen.

Hahn also invented and re-invented traditional laws and customs. As Native Commissioner for Ovamboland for three decades he was not only the major source for his superiors on Ovamboland and its culture, but also one of the foremost ethnographers of the Ovambo, in addition to being considered an authority on the botany of the region.[27] Hahn carefully guarded access to knowledge about

---

[24] NAN, NAO 19, Monthly Report Ovamboland, June 1933; Kreike, *Re-creating Eden*, pp. 69–70.

[25] NAN, NAO 49, Farewell Address H.L.P. Eedes to the Ukuanyama Council of Headmen and People Assembled at the Tribal Office at Ohanguena on 12 June 1954; NAO 18, Monthly Report Ovamboland, Jan. 1929; Kreike, *Re-creating Eden*, pp. 69–70.

[26] NAN, NAO 10, ANC to NCO, 13 May 1944.

[27] Kreike, *Re-creating Eden*, p. 120.

Ovamboland which he thought held the key to control over the territory's people. In 1928 he made an urgent plea to his superiors:

> I would respectfully like to point out that it is not advisable to allow persons, whether Professors, Scientists etc. or not, to travel about alone in Ovamboland, to take photographs of chiefs, headmen and natives, and to question them about their customs. Their actions, however innocent, will tend to eventually undermine the authority of the Officials here, who, alone, administer thousands of armed natives.[28]

Hahn's main sources in turn were his favorite headmen, as were his subordinates and the missionaries.[29] In 1935, Hahn noted in his diary: "All the Ukuambi headmen visited Ondangwa & were interviewed by Mr. Hahn re. tribal affairs, witchcraft and demarcation of borders of areas".[30]

Comparing Hahn's rich private and public papers demonstrates the extent to which he edited, streamlined and at times manipulated or even misrepresented data he had collected. In one manuscript, he added a note to himself:

> State:- 37. That they [the Ovambo laws] may be modified from time to time by the will of the chief or by the new council now controlling some of the tribes. 38. That the laws are traditional (...), 42. [that] [l]aws are seldom changed. Ovambos like other Bantus are very conservative in regard to their laws and customs. If any alterations should be necessary then these are affected [*sic*] by the chief & his councillors (or by the council of headmen where there are no chiefs).[31]

During a speech at a meeting in Windhoek in 1936, Hahn let slip the extent to which he had in effect changed 'traditional' units of administration:

> The boundaries of the Omukundas [villages] are well-known by the natives. They are seldom changed. Any school opened in an Omukunda would of course take the name of that particular Omukunda. In certain places we have already knitted three or four Omukundas together, since they have been too small for individual headmen to control.[32]

---

28 NAN, NAO 18, Monthly Report Ovamboland, Nov. 1928.

29 Kreike, *Re-creating Eden*, p. 120.

30 NAN, NAO 106, Diary NCO 1935–1938, entry 26 Aug. 1935.

31 NAN, A450, 10, Mss. by Hahn with questions and notes regarding Ovambo laws.

32 NAN, NAO 13, Conference Regarding Mission and School Sites in Ovamboland and the Okavango, Windhoek, 24–25 Nov. 1936.

In addition to promoting his favorites in the formal administrative hierarchy of headmen, Hahn also gave them a key role to play in recruiting migrant labor to supply the farms and mines of colonial Namibia south of Ovamboland, and in collecting taxes when these measures were introduced in the early 1930s; Ovamboland was the principal source of African labor for South Africa's Namibian colony.[33] Hahn also provided his Senior Headmen and the regular headmen with a new source of income by granting them a monopoly over land allocation and allowing them to charge a one-time 'transfer-fee' for existing farm plots when a new owner took over the farm after the death of the previous owner. To be granted the lifetime use of a farm, the applicant had to pay a fee to the village headman and / or the Senior Headman. In the 1930s, the fee for a farm could be as high as three pounds sterling and one head of cattle. During the 1930s and 1940s, this custom spread across Ovamboland.[34]

Decentralized despotism corrupted colonial officials and their subordinates alike. In 1935, George Tobias, the leading Anglican missionary in Ovamboland, sharply criticized Hahn's administration in a mission pamphlet and alleged that serious abuses of power were taking place in the Oukwanyama district:

> The present policy seems to be simply to preserve law and order by indirect rule through Chiefs and Headmen, without doing anything to improve the conditions of the people (...). The main argument against the work of the Missionary is that it undermines the authority of the Headmen. There may be something in this, though we do all we can to instill into our converts the duty of loyalty and obedience to the Government and to the Chiefs and Headmen. What undermines the authority of the Headmen more than anything else is Police-Boy Government. They are feared and resented by both headmen and people as an upstart bullying class, who have the ear of the white ruler as his interpreters and servants.[35]

W.E. Cawthorne, Tobias' colleague, stressed that the headman of the Onanmunamu district of Oukwanyama in eastern Ovamboland, where Cawthorne was stationed, arbitrarily ejected people—including adherents of the Anglican church—

33 NAN, A450, 12, SWA Commission: Minutes of Evidence (1935), vol. 12, session at Ukuambi, 14 Aug. 1935, evidence C.H.L. Hahn, pp. 705–713.

34 Kreike, "Architects of Nature", pp. 232–237.

35 NAN, NAO 13, Ovamboland Mission: Quarterly Paper of Thanksgiving and Intercession, No. 35, July 1935.

from their farm plots, causing at least twenty-eight families to flee the area. In addition, the 'police boys' of the Oshikango office had run rampant, extorting 'fines' from the hapless owners of any donkeys they found wandering without a herder. Because the accusations had become public, Hahn demoted the Senior Headman (but left him in charge of the area) and forced the 'police boys' to return the extorted money. Hahn's superior, the Administrator of South West Africa (Namibia) concluded that the accusations of a lawless 'police-boy government' were entirely unwarranted. Tobias' accusations were dismissed as the product of a mind on the verge of a nervous breakdown. Hahn—obviously relieved that he had got off the hook—wrote in the margins of the letter he received from the Administrator about the decision: "File & look pleasant!"[36] For the second time in his career, Hahn emerged unscathed from what could have become a major scandal. These were not isolated incidents. After Hahn's retirement in 1947, the Senior Headmen of Oukwanyama complained that they faced increasing resistance against the forced labor that they habitually had imposed to work on such projects as road building within the reserve, suggesting that forced unpaid labor had been key to Hahn's administration. Two years later, an anonymous letter accused the Senior Headmen of the Oukwanyama district of abusing their powers to make subjects work in their fields without compensation, demonstrating that forced labor did not end with the departure of Hahn.[37]

Hahn's personal rule and the regime of decentralized despotism over which he presided indirectly but deeply shaped environmental use and management in Ovamboland. Hahn made the majority of the headmen subservient ("sub-headmen" in his words) to a small coterie of Senior Headmen. The headmen controlled arable land, labor and wage labor opportunities (through their role in migrant labor recruitment), water reservoirs and modern firearms (for hunting and for defense against predators). Senior Headmen and Ovamboland's remaining

[36] NAN, NAO 13, W.E. Cawthorne (Holy Cross Mission) to ANC, Onamunama 27 Oct. 1933; statement by Cawthorne, CEM Ovamboland, 30 Aug. 1935; O / C Native Affairs Oshikango to NCO, Oshikango, 19 Sep. 1935; NCO to Sec. SWA, Ondangwa, 14 Oct. 1935; and Administrator SWA to NCO, Windhoek, sd [received Ondangwa, 19 Nov. 1935].

[37] NAN, NAO 63, Report ANC, 14 Dec. 1947; NAO 98, anonymous communication to Ohamba Nakale [Eedes], received at Ondangwa 17 Jan. 1949, and ANC to NCO, Oshikango, 20 Jan. 1949.

kings received their main income from 'selling' villages to headmen. Since they allocated villages for the lifetime of the buyer, the founding of new villages was a lucrative undertaking.

The simplest way to earn additional income was to subdivide existing villages, which was possible because the colonial administration had placed a cap of 100 on the number of farms per village in order to prevent overpopulation and deforestation. The practice of subdividing villages mushroomed in the 1940s and 1950s. Often, the colonial administration was unaware that the villages had been subdivided. In 1952, only 456 of Ondonga district's 603 villages were recorded in the administration's books.[38]

Hahn's enhancement of his chosen headmen's personal powers at the local level also meant that village forest conservation was largely up to the village headman. The colonial government's policy of the 1940s and 1950s, of preserving existing forest and bush vegetation in formal village forest reserves, met with little success. A 1941 letter by Hahn directed to his superior in Windhoek stated: "It has always been the policy of this office to encourage the protection of indigenous trees and in every 'umkunda' (area) [village] a portion is, as far as possible, always kept as bush and forest reserve. The difficulty, however, is one of control".[39] As an example, in the 1950s, Gabriel Kautwima, who had been personally selected by Hahn's successor Eedes as a Senior Headman in the Oukwanyama district, nevertheless permitted farm plots to be cleared in his Omhedi village's designated forest reserve.[40] The action might be construed as merely ironic: Hahn's and Eedes' chosen headmen using the despotic powers bestowed upon them to undermine their patrons' conservation policies. But, in fact, there was much more to it. From 1920 to 1954, Hahn and Eedes themselves not only failed to implement but actively campaigned against the introduction of South Africa's conservation regulations into Ovamboland.

---

38 NAN, NAO 100, Chief Kambonde to NCO, 8 Aug. 1952, statement of Holongo Amshelelo-nanda at Ondangwa, 4 July 1952, Chief Kambonde to NCO, Okaloko, 21 Jan. 1954, NCO to Chief Kambonde [Ondangwa], 19 Jan. 1954; NAO 51, NCO to Chief Kambonde, Ondangwa, 12 Dec. 1951, Chief Kambonde to NCO, Okaroko, 18 April 1952; NAO 104, "Ukuambi Affairs 1932–33", Iyambo Nule to Hahn, n.p., n.d., received Ondangwa, 15 Nov. 1934 (two letters), Onimnandi, Uuk-wambi, n.d. [received Ondangwa, 13 June 1934]. See also NAO 99, statement of Mingana Shikongo, 28 Oct. 1950.

39 NAN, SWAA 3, NCO to CNC, Ondangwa, 2 June 1941.

40 Joshua Mutilifa, interview by author, Omhedi, 8 March 1993.

Hahn conceded that environmental degradation had become an issue in Ovamboland since at least the early 1930s. The massive influx of refugees from Angola and the subsequent deforestation and a severe 1929–1932 famine were testimony to the soft environmental underbelly of Hahn's Ovamboland project. In public, however, Hahn denied that conservation was an immediate priority.[41] Although Hahn and his subordinates raised conservation concerns regularly at meetings with Ovambo kings and headmen, the meetings were more about political control than about the environment. In 1934, for example, Hahn met with the Senior Headmen and village headmen of the Oukwanyama district about suppressing wildfires, preserving trees and opening up eastern Ovamboland as a settlement area.[42]

In a 1941 report to his superior, Hahn conceded that the construction of heavily palisaded homesteads was "responsible for the cutting of many trees, poles, saplings and brushwood, and this cannot be stopped unless the whole system of kraal building is altered". Yet, Hahn advised strongly against the application of the regulations to control timber cutting, as issued under the Native Trust and Land Act No. 18 of 1936 in Ovamboland, for three reasons. First, he emphasized that the inhabitants of Ovamboland traditionally were allowed to cut trees freely for "domestic uses" and that "[a]ny interference with such a right would definitely, at their present state of development not be understood and lead to discontent and disturbances". Second, he argued that introducing timber fees for cutting wood to construct homesteads as occurred in South Africa was impossible because Ovamboland's economy was not sufficiently monetized. Third, he emphasized that the situation in Ovamboland differed from that in South Africa.[43] Indeed, he openly stated that he did not enforce South Africa's conservation regulations in Ovamboland because he believed that 'traditional' Ovambo conservation practices made such regulations superfluous. Eedes, in turn, considered Ovamboland to be unique, so that enforcing the regulations was inappropriate.[44]

---

41 NAN, A450, 10, "Agriculture".

42 NAN, NAO 19, Monthly Report Ovamboland, Sep. 1934.

43 NAN, SWAA 3f. Administration, Forestry: Indigenous Forests Ovambo A1 / 2 (I), NCO to CNC, Ondangwa, 2 June 1941.

44 NAN, NAO 101, NCO to CNC, Ondangwa, 17 April 1948, and Social and Economic Planning Advisory Council (van Eck), Report No. 9 (1946) (UG 40 of 1941), pp. 16–28, 50–54.

The only notable tree conservation measure that was implemented was to require the missions to obtain permits from the administration in order to cut down trees.[45] Missions and the colonial administration itself were significant wood consumers and they competed with villagers for the same species, although they preferred the larger trees that could be cut into timber. A 1931 report by the Assistant Native Commissioner noted that in the preceding decade, four large missions and 100 outstations and schools had been constructed in Oukwanyama alone, consuming at least 2,800 timber-quality trees. A single 1955 request for seven Finnish Mission stations required cutting down 1,260 mopane, 140 tamboti (*Spirotachys africana*) and 60 Lowveld cluster leaf (*Terminalia prunioides*) trees.[46] Missions supposedly used dead trees to construct churches and schools, but in reality, dead trees of the preferred species were increasingly scarce in many places by the 1950s. In the Oniipa area of Ondonga district, for example, dry tamboti was rare by the late 1950s, although in 1955 it could still be obtained in the Uukwambi district, and the FMS requested permission to cut down green Lowveld cluster leaf and mopane.[47]

Hahn's unpublished private papers contain evidence that suggests he was more concerned about the state of the environment than he publicly acknowledged. He listed overpopulation as a result of immigration from Angola, exhausted soils, deforestation and overstocking as major problems. In a manuscript that may have dated to the late 1930s or early 1940s, he appeared confident that exploring and exploiting water resources in eastern Ovamboland would make new lands available for settlement and thereby relieve population pressure. In his draft Annual Report for 1942, he emphasized that overpopulation had become a "priority issue" and he proposed to end any further immigration from Angola.[48]

---

45 NAN, BAC 131, Agricultural Officer Ovamboland to Bantu Commissioners Ondangwa and Oshikango, [Ondangwa?] 28 Jan. 1957, and FMS to Bantu Commissioner Ondangwa, Oniipa, 16 and 27 Jan. 1957.

46 NAN, SWAA 3, O/C Oshikango to NCO, Oshikango, 17 March 1931; BAC 131, Agricultural Officer Ovamboland to Bantu Commissioners Ondangwa and Oshikango, 28 Jan. 1957, and FMS to Bantu Commissioner Ondangwa, Oniipa, 16 Jan. 1957. See also Sec. SWA to Administrator SWA [Windhoek], 27 March 1955, and Superintendent FMS to NCO, Oniipa, 27 Jan. 1957.

47 NAN, BAC 131, Agricultural Officer Ovamboland to Bantu Commissioners Ondangwa and Oshikango [Ondangwa?], 28 Jan. 1957.

48 NAN, A450, 10, Mss: page marked "56", and Mss, apparently a draft of an annual report for 1942. Both these documents are part of Hahn's private papers (A 450) and not of the NAO collection which holds the files pertaining to his service as Native Commissioner.

Yet in response to inquiries from the Chief Native Commissioner in Windhoek regarding Portuguese complaints that Ovamboland subjects illegally cut the precious Transvaal teak (*Pterocarpus angolensis*) timber trees in Portuguese territory, Hahn allowed his assistant in Oukwanyama to defend the Ovambo 'tribal' record of tree conservation. The latter claimed that

> [u]nder existing tribal customs in Ovamboland the ordinary Natives require special permission from their Chief or Councillor Headman to cut down green trees. In Ukuanyama [Oukwanyama district] a Native is only allowed to cut down trees when clearing new lands and no land may be cleared without the prior authority of the Councillor Headman of the Area. Even then the cutting of large trees for clearing new fields is discouraged.[49]

Hahn conceded, however, that "[t]his law is easily enforceable in the inhabited tribal area but it becomes altogether a different question in the uninhabited Eastern Ukuanyama bush where it is impossible to have effective control over thousands of square miles", and he advised against enforcing the Portuguese law because it would lead to the exploitation of Transvaal teak trees on the Ovamboland side of the Angolan-Namibian border.[50]

The Anglican missionary W.E. Cawthorne, however, warned about the threat of environmental disaster. Cawthorne considered Ovamboland to be essentially a desert environment that could yet be saved through careful conservation efforts:

> For the present the native resources should be carefully examined. Large areas, at present too dry for settlement, could bear timber giving excellent plank wood. Some afforestation might be stimulated to aid rainfall as the country seems to be drying up (. . .). There remains the major problem that the present boundary between South West Africa and Angola cuts off a native people from their natural home and drives them to live in what is really a desert. In the nature of things, despite all palliatives, the pressure of the circumstances will become more and more severe and native development practically impossible until the unity of the people and their country is restored.[51]

---

[49] NAN, NAO 44, ANC to NCO, Oshikango, 24 March 1942.

[50] NAN, NAO 44, ANC to NCO, Oshikango, 24 March 1942.

[51] NAN, NAO 104, SWA Commission 1935, W.E. Cawthorne, Priest in Charge Holy Cross Mission Ovamboland, to the Chairman, The Constitutional Commission, Windhoek, 14 Aug. 1935.

Nonetheless, Hahn continued to oppose the introduction of any substantial conservation measures in Ovamboland beyond game conservation. In part based on his decades-long experience in Ovamboland, he genuinely may have felt that the purported threats of deforestation and desertification in Ovamboland were exaggerated and alarmist. But his hesitation about introducing conservation regulations was also politically motivated, since such measures threatened the special status that Ovamboland had gained under his personal administration, and therefore also the idiosyncratic way in which he ruled his 'fiefdom'. Hahn used the threat of disturbances as a trump card to deflect calls to bring the administration of Ovamboland in line with that of other South African reserves; implied throughout his reports is the notion that Ovamboland was stable under his "personal touch", but that it could "explode" if any radical new policies were introduced from the outside. The implementation of a comprehensive conservation policy would require abandoning his personal and indirect style of rule for a more direct interference in how the environment was managed in which scientific expertise would count more than the ethnographic and experiential knowledge that was the foundation of his authority. On a sheet in his personal files that probably dates from the 1940s, Hahn wrote: "It [development without overpopulation and environmental stress] can only be done scientifically under proper European supervision & this means a [*sic*] increased native affairs staff".[52]

The colonial conquest of Nature: Direct environmental rule

The 1947–1950 foot and mouth affair highlighted the transition from indirect environmental rule to direct interference in Nature through a second colonial conquest. The affair illustrates how the political ecology framework may serve to identify multiple struggles that affected one another without leading to a single and unambiguous environmental outcome; in fact, different measures and actions had contradictory consequences. Professional rivalries that had nothing to do with environmental concerns were important factors given the extent to which

[52] NAN, A450, 10, Mss: page marked "56".

the Ovamboland administration relied on personal rule and decentralized despotism. Native Commissioner Harold Eedes deliberately used the foot and mouth scare to introduce draconian measures in an attempt to change fundamentally cattle use and management in Ovamboland. Like Hahn, he saw himself as the king or paramount chief of the area. The 1926 monthly report for Ovamboland noted that Eedes "played" the role of chief of Oukwanyama, and after he succeeded Hahn, subjects often addressed him as *ohamba* (king) in their correspondence.[53] Much of his motivation arose from his attempt to best his predecessor Hahn and to confirm his own reputation as a can-do modernizer. The whole affair could have backfired on Eedes: first, his extreme measures (including shooting cattle) proved entirely ineffective; and second, alarm about an outbreak of foot and mouth proved unfounded not just once, but twice. Yet, Eedes turned each fiasco around and was able to strengthen his own position vis-à-vis a new generation of scientifically trained colonial experts as well as versus the headmen and population of Ovamboland. He implicitly prevented science from becoming the dominant discourse of knowledge (and power) during his tenure even as he used science to shore up his rule and status. He did not need to spell out the limitations of science: scientific knowledge clearly demonstrated its limitations outside of the laboratory. During each of the two disease outbreaks, the veterinarians concurred with the diagnosis of foot and mouth, only to discover months later when they investigated infected animals that their initial diagnosis had been wrong. Eedes explicitly blamed Ovamboland's headmen and the population at large for the outbreaks because they not only had ignored his advice to quarantine their cattle, but also had broken their promise to cooperate. Moreover, the veterinarians had made their initial diagnosis based on descriptions of the symptoms of the infected animals by Ovamboland's headmen and cattle holders. Thus in one fell swoop, Eedes discredited Ovamboland's headmen and population as minor partners in an indirect system of political and environmental governance. They had disobeyed orders and lied and proved themselves to be incapable of effectively managing their own natural resources: their indigenous knowledge base fell short not only in identifying, but also in describing the disease.

[53] NAN, NAO 18, Monthly Report Ovamboland, Sep. 1926, and NAO 98, anonymous communication to Ohamba Nakale [Eedes], received at Ondangwa 17 Jan. 1949.

During Hahn's long tenure in Ovamboland, the local population's 'primitive' but natural acceptance of Nature's plagues had justified the radical measure of cattle containment which cut off the region's livestock from outside markets.[54] Containment imposed new limits on the practice of cattle transhumance. But Hahn had refrained from any radical interventions in cattle use and management within the reserve and had come to terms with the continued movement of cattle, beef and hides across the border with Angola. Hahn successfully fought off challenges to his hands-off cattle policy in the 1920s and 1930s when he canceled cattle vaccination projects, and again in 1946, when foot and mouth surfaced across the border in Angola. His motivation in the 1920s might have been simply to keep any potential competing colonial source of authority out of 'his' reserve. An outbreak of anthrax and blackquarter evil on the border brought the colonial Chief Veterinary Officer to Ovamboland in September 1926 to inspect livestock and consult with his Portuguese counterparts. But Hahn reported in November 1926 that the presence of the officer in Ovamboland was no longer required since "[t]he question of inoculating Zone cattle which was his main duty has now ceased owing to the lack of keenness and disinclination on the part of the natives to have their stock treated".[55]

Foot and mouth made its first appearance in southern Angola in the 1930s, but it did not cross the colonial border into Ovamboland.[56] The 1946 outbreak in the Portuguese-ruled northern Ovambo floodplain, however, jolted South Africa's administration into action. The administration sent veterinarian Dr. Zschokke, accompanied by a police major, to southern Angola to investigate, and hastily seconded a veterinary officer to Hahn's staff. After his visit to Angola, Dr. Zschokke self-confidently stated: "Everything possible is being done here [in Angola] to check the spread of the disease into Ovamboland and the natives have all been instructed to report immediately any causes (...). *Foot and Mouth disease is unknown to them but the symptoms have been well described*".[57] The veterinary officer seconded to Hahn's staff recommended the immediate implementation of a five-mile-wide stock-free zone along the Angolan border. Any trespassing

---

[54] See chapter 6.
[55] NAN, NAO 18, Monthly Reports Ovamboland, Sep. and Nov. 1926.
[56] NAN, NAO 20, Annual Report Ovamboland 1937.
[57] NAN, NAO 21, Quarterly Report Ovamboland, Jan. – March 1946. Emphasis added.

cattle would be shot on sight, and any cattle products would be impounded. Hahn overruled the veterinarian, arguing to his superiors that such measures to close the border "are out of the question in the near future in that part of the country, not only because the natives would not understand them and would not carry out our instructions" but also because the Portuguese would have to fully cooperate. Instead, a veterinary cordon with a patrolled stock-free zone was located in the far south of Ovamboland, near the Etosha Pan, to prevent the infection of animals further south.[58]

Reports that the disease had spread into northwestern Ovamboland in February 1946 could only be investigated in August because heavy flooding during and after the rainy season made the area inaccessible. Careful inspection of the affected animals revealed that the lesions to the hooves and mouths that the inhabitants in the border region had described, and that veterinarians and officials had interpreted as a sure sign of foot and mouth, had different causes. The mouth lesions had been caused by the rather innocuous lumpy skin disease virus and those on the hoofs were the result of Crotolaria plant poisoning. Further investigation revealed isolated signs of recent lumpy skin infection along the entire border with Angola as well as on the other side, leading the officials to conclude that the disease had originated in Angola.[59]

When Eedes formally took over from Hahn in January 1947, he immediately set the stage for a more interventionist administration. In his first quarterly report he stated that Ovamboland was overpopulated and overstocked and that it was critical to develop water resources in the large unused areas outside of the floodplain. One of his first actions was to fence the administration's compound at Ondangwa, a measure that can be seen as symbolic of his new policy of direct environmental rule. In June 1947, he hosted the Assistant Director of Native Agriculture of the Union of South Africa on a visit to Ovamboland "for the purpose of investigating the necessity of the establishment of a local Agricultural Section". Eedes also moved the 1946 foot and mouth cordon in the far south of Ovamboland even further south to Namutoni, effectively establishing what

58 NAN, NAO 15, GVO Ovamboland to Director Agriculture, 18 Jan. 1946, "Investigation: Foot and Mouth Disease"; and NAO 21, Quarterly Reports Ovamboland, Jan. – June 1946.

59 NAN, NAO 59, GVO Grootfontein, Annual Report 1947, and NAO 21, Quarterly Report Ovamboland, Jan. – March 1946.

later came to be known as the Red Line.[60] Although the veterinary cordon in its new position no longer interfered with cross-border cattle movements, it now cut across the routes from Ovamboland to the salt pans on the edge of Etosha Pan. Under pressure from King Ushona Shimi of the Ongandjera district, Eedes requested special permission from his superior in Windhoek to allow salt gathering across the veterinary cordon. The Native Commissioner's telegram read:

> Natives of North Western tribes, particularly the Ongandjeras, complain bitterly against Foot and Mouth Disease restriction which prevents them crossing line to obtain salt from pans which strictly speaking are in tribal areas (...), Natives cannot exist without sufficient supplies of salt (...), recommend Natives be allowed proceed salt pans with donkeys only (...), No pack oxen will be taken.[61]

Despite the complications it caused, Eedes maintained the veterinary cordon even though almost a year had gone by without any new reports of foot and mouth by the headmen. But Eedes had little confidence in the chiefs and headmen that Hahn had relied on for his information, and, moreover, his veterinary officer claimed that the disease was unknown in Ovamboland, suggesting to Eedes that the local population would be incapable of recognizing the disease even if they saw it.[62] As a matter of fact, early in his tenure as Native Commissioner, Eedes distanced himself from the headmen who had been the closest allies to Hahn. Immediately upon succeeding Hahn in 1947, Eedes publicly humiliated Nehemia Shovaleka, one of the most powerful headmen of Ovamboland, by emphasizing that the Senior Headman of the Oukwanyama district was a mere "native interpreter" (and "police boy") who owed his high position solely to the Native Commissioner. In a statement to the Oukwanyama Council of Headmen, Eedes exhorted:

60 NAN, NAO 60, Quarterly Reports Ovamboland, Jan.-Sep. 1947.

61 NAN, NAO 58, Telegram (E179) NCO to Secretary Windhoek [Oct. 1947], and Chief Ushona Shimi [to NCO?], Ongandjera, 24 Oct. 1947.

62 NAN, NAO 60, Quarterly Reports Ovamboland, Jan.-Sep. 1947; NAO 15, Veterinary Officer Ovamboland (Zschokke) to Director Agriculture, Investigation: Foot and Mouth Disease, 18 Jan. 1946; and NAO 21, Quarterly Reports Ovamboland, Jan. – June 1946.

> I wish to record further that the mover of the resolution [Nehemia Shovaleka], although a Headman, is actually a Native interpreter. As such he has access to information, and is, on occasion, able to overhear conversations between officials, and other Europeans. He has had a lot of attention paid to him by a certain class of visitor, with the result that he has now rather an inflated opinion of himself. Although he is not respected by all the Ukuanyama Natives, he is feared because of his official position as Interpreter. He is very well known to me, as he commenced his service with me at Namakunde many years ago.[63]

Two years later, Eedes limited Nehemia Shovaleka's power over the Oukwanyama district by approving moving its Tribal Council seat (and the office of the tribal secretary to the council) from Nehemia's Omhedi homestead to a location at Omafo in a neighboring subdistrict: "The [Oukwanyama] Headmen intimated that Nehemia's occupation of the Tribal kraal had given cause for a certain amount of ill-feeling and a suspicion of favouritism and their resolution is therefore supported".[64]

Eedes returned to Ovamboland with a dream: to eradicate stock diseases in Ovamboland as he had done in Okavango in the late 1930s, and he was willing to use all means necessary to accomplish his goal. Upset that Ovamboland had been abandoned to livestock diseases, Eedes believed that Ovamboland could be made disease-free fairly easily if a 'free zone' was created along the entire border with Angola. His plan entailed transforming a five-to-ten-mile-wide zone along the Angolan-Namibian border into a stock- and settlement-free buffer to prevent re-infection across the border. The plan required the forced removal of up to 20,000 people from the border districts of Oukwanyama and Ombalantu, but Eedes reasoned that the displaced could easily be accommodated in eastern Ovamboland once a water infrastructure had been put into place. He shared his brainchild confidentially with a Mr. Randall of the Church of England Mission "and told him that it was quite likely that I [Eedes] would not be supported by the Administration. He unfortunately reported my conversation to Mr. Allen [the Administrator for South West Africa], who asked me to submit a report". Eedes wrote the Chief Native Commissioner that "[t]he matter has not been discussed

[63] NAN, NAO 63, NCO to ANC, Ondangwa, 15 Nov. 1947.

[64] NAN, NAO 51, ANC to NCO, Oshikango, 12 Feb. 1949. Eedes noted at the foot of the letter that he approved of the proposal to move the tribal seat.

with the Natives, or with the Portuguese Authorities, and for reasons of which you are aware, it is suggested that it be kept strictly confidential for the present".[65] But Eedes' request for secrecy came too late, because several weeks earlier, in June 1947, his boss had already informed George Tobias, by now Bishop in Windhoek, that the administration was studying a plan to create a ten-mile buffer zone along the border and to relocate the local residents.[66]

The missionaries' reaction was blistering. Early in 1948, Tobias wrote to the Additional Native Commissioner in Windhoek that removing the Oukwanyama population would be "a disastrous mistake" and expressed his belief that "the Administration will never sanction such an arbitrary and despotic act".[67] The Anglican missionary in Oukwanyama, the Reverend Dymond, urged caution:

> There is a vast amount of disquiet and of threats of resistance to evacuation going on among the ova Kuanjama (...). The general theme is, 'I will die—I will be shot in my kraal—before I will consent to leave it' (...). There is general scepticism about the existence of 'cattle-disease'; and, if cattle-disease can be demonstrably proved to exist, there is no sympathy whatever with the proposed method of its cure—the evacuation of fertile territory, the best in Ovamboland (...). A lack of confidence in the European Administration is growing and will certainly increase as long as the people have reason to believe that plans are afoot to remove them from their homes for purposes of which they are ignorant.[68]

The senior FMS missionary in Oukwanyama, the Reverend Bjorklund, was even blunter:

> The man in whose brain such an idea has matured must be either an abnormal man or else an enemy of the Ovambos and of the Missions (...). If it [the buffer zone] is to be isolated only for pure nonsense like cattle-disease, there will be so great disturbances that the Administration will lose all Ovamboland; and all Ovambos from the mines, the farms and the towns will leave their work and run back to Ovamboland their home-country (...). Disturbances may perhaps happen as a result of the cutting of the Oukuanjama. Further the Reds whose

---

65 NAN, NAO 59, NCO to CNC, 2 July 1947.

66 NAN, NAO 59, CNC to Bishop of Damaraland (Tobias), 13 June 1947.

67 NAN, NAO 59, Bishop of Damaraland to Additional Native Commissioner, Windhoek, 12 Feb. 1948.

68 NAN, NAO 59, Dymond to the Bishop of Damaraland, 27 Jan. 1948, appendix to Bishop of Damaraland to Additional Native Commissioner, Windhoek, 12 Feb. 1948.

> eyes are everywhere—know what is going on in this country as well as other countries, and they perhaps will come on the pretext of liberating the Ovambos from a tyrannical Administration.[69]

The missionaries' warnings echoed those of Hahn against interfering directly in how the inhabitants of Ovamboland raised their cattle: that to do so would trigger resistance. Eedes' worried superiors in Windhoek asked him to investigate, but, not surprisingly, the Native Commissioner and his staff found no evidence of any unrest. On the pretext of collecting the data for a census, Assistant Native Commissioner C.S. Holdt, who was stationed in Oshikango, Oukwanyama, had been quietly assessing how many people would be affected by a forced removal since early December 1947. Holdt believed it to be likely that most of the district's inhabitants had heard rumors about the removals and that there was a great deal of distrust of the administration. Holdt advised Eedes that "[i]t is doubtful whether the people would voluntarily accept such a scheme. There are many reasons for this. But it could nevertheless be put to them. There appears to be no reason for not doing so, especially now. If they do not accept it, then the Administration will have to consider whether they are prepared to put through the scheme even in the face of opposition".[70]

Eedes disagreed with his subordinate about the need to consult the Oukwanyama leaders; he felt that it was too soon. Fearing that the administration might prematurely vote down the plan because the missionaries "may have sabotaged the efforts of the Administration to develop Ovamboland", Eedes suggested that Dymond and Bjorklund be warned to stay out of politics or face expulsion from the reserve. Ultimately, in a letter marked confidential, Eedes proposed to shelve the plan for future consideration and added: "I suggested the present scheme, which the Administration told Bishop Tobias was a good one, and which has the support of the Veterinary and Police Authorities. I am quite prepared, however (…) [to] tell the Natives that the scheme has been abandoned. It must be borne in mind that the Portuguese Authorities may adopt a similar scheme on their side of the border, and that they would not consider the wishes of the Natives

[69] NAN, NAO 59, Bjorklund (FMS) to Dymond, undated, appendix to Bishop of Damaraland to Additional Native Commissioner, Windhoek, 12 Feb. 1948.
[70] NAN, NAO 59, ANC to NCO, 21 Feb. 1948.

(…). There is also always the possibility that the Portuguese Authorities will suddenly one day prohibit the continual movement of Ukuaanyama cattle in and out of Angola".[71]

Windhoek accepted the recommendation. Eedes reported that on 6 March 1948, he addressed the Oukwanyama district Tribal Council meeting as follows:

> Reports have been made to the Government in Windhoek that there is a very unpleasant atmosphere of suspicion everywhere, and that there is a lack of confidence in the European Administration. The Government has instructed me to give you the following message:—There was never any proposal on the part of the Administration to evacuate the Ukuanyama (…). He [the Administrator] hopes that the reports that you are disturbed by these rumors are not true. As you know we have talked for many years about the possibility of eradicating stock disease in Ovamboland (…). If Ovamboland was free from stock disease, the possibility was foreseen of Ovamboland being reinfected from Angola in the absence of a proper "line" between the countries. I wish to tell you now that there is no intention of evacuating the Ukuanyama tribe, and that even a suggestion, made by somebody, of a small evacuated strip along the border line was never entertained by the Chief native Commissioner, or by the Administrator.[72]

The incredulous Eedes could hardly believe his ears when the Oukwanyama headmen responded to his hypocrisy in kind. Their spokesperson said: "We have never heard about this matter until now. After having heard your words we are now afraid (…). Who gave the reports to the Government in Windhoek? (…) How does 'Nakale' (Mr. Eedes) know about any unrest if we do not". Eedes advised his superior against publicly identifying Dymond as one of the culpable, because "he is quite likely to show his Converts and Native Priest, your letters (…) addressed to Bishop Tobias" about the removal plans and embarrass the administration.[73]

Eedes' ally, the veterinarian Dr. Zschokke, did his part to keep Eedes' plan alive. In October 1948, seven months after Eedes' ruse of publicly denying his own plan, Dr. Zschokke confidentially urged his departmental superior in Windhoek, Director of Agriculture Dr. J.S. Watt, to close the Angolan-Namibian border to

---

71 NAN, NAO 59, NCO to CNC, 25 Feb. 1948.
72 NAN, NAO 59, NCO to CNC, 8 March 1948.
73 NAN, NAO 59, NCO to CNC, 8 March 1948.

cattle movements with a fence. Zschokke's letter was critical of what he portrayed as the old unenlightened Ovamboland establishment (Hahn and the missionaries) and praised Eedes as an ally of rational and energetic scientist-experts (like himself and Watt) who could and should shake up the dozing bureaucrats in Windhoek:

> I am writing this letter to you, because I feel that the mere fact that the Administration has sort of shelved a problem, of utmost important [*sic*] in my opinion, does not justify any closing of eyes on our part. Mr. Eedes closed the northern boundary [with Angola] for to and fro movements of stock in the Okavango Territory after the first lungsickness campaign in 1938. We both know only too well, what was benefitted by this action (...). The border-line of Ovamboland was never closed, except for a short period in 1946, when we had to take over control, on account of F&M [foot and mouth]. We both know of the numerous outbreaks of various notifiable stock-diseases occurring in this part of the country (...). Mr. Eedes, then in the Okavango, was for clear territorial segregation, because he believed it to be beneficial for both countries, Major Hahn was against it for a number of reasons, not all known to me and declared the establishing of such boundary conditions (...) an impossibility, as far as Ovamboland was concerned (...). During the 1946-F&M-Campaign the closing of the border between Angola and Ovamboland, was an essentiality. Against all belief of the Ovambolanders (Missionaries and Officials) we had a stretch of 120 miles of border-line under control within a few weeks and all animal and human traffic was efficiently stopped (...). And I am more than ever before convinced that the most important factor in our endeavours to free the N[orthern] N[ative] T[erritorie]s from certain important notifiable diseases is a *frontier towards north* [*sic*], *which can be efficiently controlled.* Owing to various unfortunate circumstances Mr. Eedes proposal viz. to create an empty buffer-zone, met with disapproval on the part of the Administration. (I have made similar proposals in 1946, see my reports) (...) I should feel very happy indeed if through this letter and your subsequent action this whole complex of problems could be re-considered by the Administration in order that we are not coming out second best with our "noble allies" [the Portuguese].[74]

Zschokke and Eedes continued preparations for their project to eradicate livestock diseases in Ovamboland. In the same month that Zschokke made his plea to his boss, he successfully trained 122 volunteers from the eastern Ovamboland districts of Oukwanyama, Ondonga and Uukwambi as cattle inoculators. The

[74] NAN, NAO 59, Dr. Zschokke to Director Agriculture, Windhoek, 26 Oct. 1948. Emphasis in original.

veterinarian noted with satisfaction that the volunteers "know how to operate a syringe in a professional manner".[75] In September 1949, Eedes' assistant at Oshikango triumphantly reported that the

> [Oukwanyama] Headmen unanimously agree to any type of inoculation (...) and ask the Administration be requested to make as early a start as possible and that cattle diseases be tackled in the following order of priority: (1) lungsickness, (2) Anthrax and (3) Blackquarter. Headman Nehemia points out that the best months for an inoculation campaign are April to July as, during that period, most of the cattle are back from the cattle posts.[76]

In early November of the same year, the chiefs and headmen of western Ovamboland also agreed to cooperate with the vaccination program, which included the use of branding to mark vaccinated cattle. The Ovamboland veterinarian wrote to his superior in Windhoek:

> Obviously there was and still is a certain amount of antagonism amongst the older people against such novelties (...). But the fact that the younger generations agreed to come together (...) and discuss all questions and proposals (...) resulting eventually in a clear decision to brand and inoculate in future, seems to indicate that a step towards betterment of conditions was achieved.[77]

During the same month, the Oukwanyama headmen reiterated their commitment to the vaccination project and agreed that vaccinated cattle would be branded, and the veterinarian services trained another sixty-four inoculators from the districts of western Ovamboland. Eedes' prediction that the Portuguese one day would simply close the border became a reality at the end of 1949 as the Angolan veterinary services prepared for a grand effort to eradicate the principal cattle diseases on their side. No doubt also to put pressure on his superiors, Eedes informed Windhoek that until all Ovamboland's cattle had been certified as vaccinated, the Portuguese would no longer allow any livestock from south of the border to enter Angola beginning on 2 January 1950.[78] Thus, three years after

[75] NAN, NAO 59, Zschokke to Director Agriculture Windhoek, 5 Nov. 1948.
[76] NAN, NAO 59, ANC to NCO, 17 Sep. 1949.
[77] NAN, NAO 59, Zschokke to Director Agriculture Windhoek, Ombarantu, 4 Nov. 1949.
[78] NAN, NAO 59, ANC to NCO, 9 Nov. 1949; Zschokke, Training-Course of Inoculators: Ombarantu, Nov. 3–6, 1949; Zschokke to Director of Agriculture Windhoek, Oshikango, 0 [*sic*] Nov. 1949; NCO to CNC Windhoek, 21 Jan. 1950; telegram NCO to Sec. SWA, 29 Dec. 1949.

Eedes took over from Hahn, and despite setbacks, the stage seemed finally set for Eedes to eradicate cattle diseases in Ovamboland.

Eedes must have felt as if Providence was on his side because a renewed foot and mouth outbreak—this time in the Okavango Native Territory due east of Ovamboland—added a sense of urgency, justifying radical and rapid action. Claiming that he suspected that Angola was the source of the infection, Eedes immediately prohibited any movement of cattle and cattle products across the Angolan border east of Oshikango in either direction, effectively preventing the return of large numbers of cattle from Ovamboland that had been at cattle posts in Angola—locations beyond the floodplain where the cattle were kept during the dry season only. Eedes personally attended a Tribal Council meeting in Ovamboland's Oukwanyama border district and warned the headmen that their herdsmen should not cross back from Angola into Ovamboland with their cattle until at least March. Eedes promised to send twenty-eight police officers with vehicles to patrol the border after the leading Senior Headman, Vilho Weyulu, expressed fears that the herdsmen might not obey him and his colleagues. The Native Commissioner threatened that any cattle found crossing the border could be shot, and added:

> The Government has also decided to put a fence of wire (...) along the border line (...) there will be only one gate and this will be at Oshikango. No stock will be able to get through this fence. Any person climbing through this fence or cutting it or damaging it in any way may be punished. These measures are being taken by the Government to protect Ovamboland from stock disease. They are necessary as I am sure you will realise. I hope, therefore, you will co-operate with the Government.[79]

Eedes reported that the headmen of Oukwanyama agreed with his proposals: "After discussion the Principal headmen, the sub-Headmen and people present agreed to co-operate with the Government. They asked that the question of the return of the cattle from Angola be decided by the end of March next".[80] In a generous mood because of the ready cooperation he received from the headmen, the Native Commissioner—after initially refusing to intervene—subsequently

---

[79] NAN, NAO 58, NCO to the Headmen and People of Ukuanyama, 10 Jan. 1950.

[80] NAN, NAO 58, NCO to the Headmen and People of Ukuanyama, 10 Jan. 1950.

successfully negotiated with his Portuguese colleagues for the release of a herd of 180 cattle that was owned by the village headman Panguashimi Nehadi from Namibia's Oukwanyama district. Eedes supplied the headman with a pass to cross into Angola and collect his cattle from Portuguese custody. Handing him the pass, Eedes went out of his way to remind the headman not to bring his cattle across the border but to temporarily "place the cattle with relatives and friends in Angola" instead. Totally disregarding Eedes' advice, Panguashimi Nehadi smuggled his cattle back into Ovamboland. The Secretary for South West Africa, determined to set an example, ordered Panguashimi's cattle to be shot, and instructed Eedes in future to shoot one or two head from each herd stopped at the border.[81] Eedes noted in his diary that he traveled to the Oshikango border post on 13 February to shoot cattle suspected of being infected with foot and mouth; the cattle probably belonged to Panguashimi.[82]

But the very next day Eedes had a change of heart. He sent an urgent telegram to Windhoek proposing to cease the cattle killings because he found that foot and mouth already had spread deep into Ovamboland.[83] Reluctantly, since it meant that he had to shelve his plan for a border barrier once more, Eedes recommended that the foot and mouth cordon be withdrawn from the Angolan border to the far south of Ovamboland, where it had been situated during the 1946–1947 outbreak. A disillusioned Eedes ended his telegram by placing the blame for the failure to contain foot and mouth—and of his plan to eradicate stock diseases—squarely on "the Ovambo" in general:

> The Ovambos have definitely shown that they are not prepared to co-operate and that they will defy instructions given to them for their own good and for the protection of South West Africa as a whole Stop It is clear that other natives not yet detected in addition to Panguashimi introducing infected herds from time to time.[84]

---

[81] NAN, NAO 59, ANC Tsumeb to NCO Ondangwa, 9 Jan. 1950, and NCO to CNC, 3 Jan. 1950; NAO 58, NCO to Doctor Schatz (GVO Ovamboland), 11 Feb. 1950; NCO to Commander SA Police Cordon Ovamboland, and NCO to CNC, 12 Jan. 1950.

[82] NAN, NAO 106, Diary NCO 1949–1954, entry 13 Feb. 1950.

[83] NAN, NAO 58, Telegram NCO to Sec. SWA, 14 Feb. 1950.

[84] NAN, NAO 58, Telegrams NCO to Sec. SWA, 14 Feb. 1950, and NCO to Doctor Schatz (GVO Ovamboland) and to Noncommissioned Officer SA Police Cordon Ovamboland, 11 Feb. 1950, and NCO to CNC, 12 Jan. 1950.

The policemen withdrew to what became the Red Line and the southern border of Ovamboland.[85] Several months later, the veterinarian inspected a herd in Oukwanyama that had been suspected to be infected with foot and mouth. His conclusion was startling: "Dr. Zschokke (…) came to an immediate conclusion that there was no Foot and Mouth disease in Ovamboland—in fact he said that there never was!"[86] In July, without sharing Dr. Zschokke's revelation, Eedes issued a stern warning to Ovamboland's chiefs and headmen:

> it is the duty of all men to report immediately any sickness in his cattle to his Headman. It is then the duty of the Headmen to report it to me (…). I wish to warn you that failure to obey these instructions will necessitate strong action being taken by me against the persons who fail to make reports to me.[87]

Only in November 1950 did Eedes inform the headmen of the border districts of Ombalantu and Oukwanyama and the interior Ondonga district that cattle could be brought back to Ovamboland from Angola, a full four months after he had alerted his superiors in Windhoek that the foot and mouth alarm—the rationale for the prohibition—had been erroneous.[88]

Eedes may have manipulated the foot and mouth scare in part to get even with Hahn, his long-time rival who also claimed to be the leading expert on Ovamboland. While serving as Native Commissioner in 1938 in Okavango after he had been removed from Ovamboland by Hahn, Eedes had eradicated lungsickness through a successful vaccination campaign.[89] Native Commissioner Hahn was unwilling and unable to repeat his disgraced subordinate's accomplishment in Ovamboland. Hahn canceled two planned vaccination campaigns in Ovamboland at the last moment and openly expressed doubts about both the political and economic benefits of lungsickness vaccination there. The feud was far from over in 1947, when Eedes succeeded Hahn while foot and mouth threatened Ovamboland. In his first report as Native Commissioner there, Eedes quoted

[85] NAN, NAO 60, Quarterly Reports Ovamboland, Jan. – June 1950.

[86] NAN, NAO 60, Quarterly Reports Ovamboland, Jan. – June 1950.

[87] NAN, NAO 59, NCO to all Ovamboland Chiefs and Headmen, 29 July 1947.

[88] NAN, NAO 59, Kaibi Mundjele to NCO, Ombalantu, 2 May 1950; NAO 60, Quarterly Report Ovamboland, Oct. – Dec. 1950; NAO 59, ANC to Kwanyama Headmen, 29 Nov. 1950; NCO to Chief Kambonde, 29 Nov. 1950; and NCO to Ombalantu Headmen, 9 May 1950.

[89] NAN, AGR 25, Senior Veterinary Surgeon to Sec. SWA., Windhoek, Nov. 13, 1941.

from a letter he had received from the headmen of the western Uukwaluthi district expressing gratitude to the government for sending "a Master whom we know" in order to dispel as "untruthful (. . .) the statement made by a certain Ovamboland official to the effect that my previous experience in Ovamboland was limited to the Ukuanayama [Oukwanyama] and Ombalantu tribal areas".[90] The official Eedes referred to was his predecessor Hahn.

Using foot and mouth as an opportunity to best Hahn on his own Ovamboland turf and to prove his own prowess, Eedes gambled not only with his own reputation, but also that of the colonial veterinary service, which had not had a permanent presence in Ovamboland before the foot and mouth affair. Even more tragic and longer-lasting in its effects, Eedes' actions put the livelihoods of floodplain's cattle holders at stake. Veterinary science in Ovamboland, Native Commissioner Eedes and Ovamboland cattle holders all suffered, but they paid very different prices. The institutional introduction of veterinary science in Ovamboland was tarnished and the territory's first veterinary officer, Dr. Zschokke, was left with little professional integrity because he had allowed Eedes to run roughshod over him. Dr. Zschokke was clearly not a scientist who subverted colonialism, to the contrary. Eedes' reputation was dented, but he remained the Native Commissioner until his retirement in 1954, re-emerging from the foot and mouth controversies as the uncontested source of authority in Ovamboland in political and environmental matters. Along with the neighboring Kaokoland and Okavango Reserves, Ovamboland retained its unique status as a Prohibited Area under Proclamation No. 26 of 1928. Entry into Ovamboland required a special permit from the Secretary of South West Africa and the reserve remained under the control of the Chief Native Commissioner in Windhoek and the Native Commissioner in Ondangwa. Except in 'technical matters', members of other departments, including scientifically trained experts in the environmental realm, namely veterinarians and agricultural officers, were wholly subordinate to the Native Commissioner.[91]

Ovamboland's inhabitants, however, paid the heaviest price, because the foot and mouth veterinary barrier remained in place as the Red Line long after Eedes had retired. Moreover, the foot and mouth affair contributed greatly to a colonial

90 NAN, NAO 60, Quarterly Report Ovamboland, Jan. – March 1947.
91 NAN, NAO 62, Permits: General, NCO Ondangwa, Sep. 14, 1952.

re-evaluation of Ovamboland's cattle from a subsistence resource into a health, and, subsequently, an environmental threat.[92] The affair also entrenched the perception that cattle holders and indeed 'the Ovambos' as a group were not the noble and self-sufficient savages that Hahn sometimes had described them as, but rather obstinate primitives who needed to be dragged into the modern world kicking and screaming, if necessary. Achieving modernity therefore would require full colonial control over Ovamboland's population, resources and environment and the firm exercise of power.

The foot and mouth affair also highlights a particular environmental twist to Mamdani's point about the autocratic nature of colonial rule. Highly authoritarian and fairly independent Native Commissioners made policy choices and implemented policies affecting environmental resources that were strongly motivated in part by personal rivalries. The affair also speaks to the concept of the second colonial conquest as a conquest of Nature. Hahn implemented the indirect policy of containing cattle diseases by confining the animals to the reserve. His successor Eedes attempted to intervene directly in cattle management and use and to eradicate cattle disease, even though the goal remained elusive. The impact of the political ecology of insecurity and the repercussions of successive regimes of indirect and direct environmental rule had dramatic consequences in north-central Namibia and elsewhere in the colonized world.

---

[92] See chapter 6.

4

Fierce species: Biological imperialism

The concepts of biological or ecological imperialism and biological exchanges highlight the role of biological invaders in environmental change. The European conquest of the Americas, for example, was as much a product of the (unintentional) introduction of Old World biological species as of military might. Biological invaders from the Old World, including smallpox germs, sheep, cattle, horses and a host of plants, accompanied European conquerors, decimating New World indigenous human, animal and plant populations, destroying the local natural environment, and transforming the Americas into a Neo-Europe.[1]

As part of the Old World, Africa has long been regarded as immune from biological imperialism from Europe. In addition, Africa was seen—implicitly or explicitly—as being complicit in the Old World's biological imperialism in the New World because several of the invading species were carried on the ships that brought African slaves to the New World—for example, malaria, yellow fever and trypanosomiasis.[2] Not only was Africa considered immune to European biological imperialism, but tropical Africa's own particular disease environment

---

[1] Crosby, *The Columbian Exchange* and *Ecological Imperialism*.

[2] Kiple, *The Caribbean Slave*.

was portrayed as a formidable obstacle to European intrusions, as illustrated by its reputation as the 'White Man's Grave'. A range of diseases, including malaria, sleeping sickness, schistosomiasis, bilharzia and river blindness, mired European military, political, economic and ecological conquest.[3]

Various Eurasian and American species introduced in Africa, however, behaved much like invasive species. Like smallpox in the Americas, bovine pleuronomia (lungsickness or cattle TB) and rinderpest (cattle plague), spread like wildfire when they were introduced in the second half of the nineteenth century, decimating domestic cattle and other susceptible domestic and wild animal species in Africa. The impact of lungsickness utterly destroyed South Africa's Xhosa society in the 1850s.[4] Moreover, European biological and nonbiological imperialism also unleashed indigenous species from their 'natural' niches in Africa and transformed them into plagues and pests, vermin and weeds as well as unintended collaborators with the colonial invaders because they disturbed precontact ecosystems. One example is sleeping sickness in Africa, an endemic disease that turned into a deadly epidemic during the era of colonial conquest.[5]

Thus, the notion of biological imperialism appears to be more appropriate to Africa than sometimes has been acknowledged. In north-central Namibia, the biological invaders included microbes that introduced at least three major human diseases (influenza, measles and the plague) and three animal diseases (lungsickness, rinderpest and foot and mouth), as well as two new animal species (the horse and the donkey). In addition, during the colonial era, several species of animals were (re)classified as vermin. Indigenous species that under the new circumstances had opportunistically infested the environment were recast as both the cause and the effect of new ecological and biological imbalances that were associated with imperialism and colonialism, including such wild animals as lions, leopards, cheetahs, wild dogs and elephants, as well as domestic cattle and goats.

---

[3] Curtin, *Disease and Empire.*

[4] Van Onselen, "Reactions to Rinderpest in Southern Africa"; Peires, *The Dead Will Arise.*

[5] Lyons, *The Colonial Disease*; Headrick, *Colonialism, Health and Illness in French Equatorial Africa*; Iliffe, *A Modern History of Tanganyika*, pp. 164–165; Kjekhus, *Ecology Control and Economic Development in East African History*; Suret-Canale, *Afrique Noire*, pp. 51, 498–499; Hartwig and Patterson, *Disease in African History.*

Invading microbes and virgin soil epizootics

The era of European imperialism witnessed the introduction of a variety of new diseases in Ovamboland. Three of the diseases affected animals, especially domestic cattle: lungsickness entered the region in the 1860s, rinderpest followed in the 1890s, and foot and mouth reached southern Africa in the mid-twentieth century.

The increased movements of people and animals through trade, hunting, exploration, missionary work and raiding throughout late nineteenth-century southern Africa facilitated the introduction of new animal diseases in the Ovambo floodplain. After killing hundreds of thousands of cattle in Europe, lungsickness reached South Africa in 1853 with a shipment of imported Frisian bulls. Transport oxen spread the disease across the region and by 1876 it had infected cattle in Ongandjera in the southern Ovambo floodplain.[6] Lungsickness remained endemic thereafter and had a significant long-term impact on land use in the region.

Far more destructive in the short term was the 1896 introduction of rinderpest in southern Africa. Rinderpest reached the Ovambo floodplain via South Africa in 1897, and decimated alike domestic livestock herds (cattle and goats) and wildlife (all single-thumb hoofed animals including antelopes were affected). Migratory herds of antelopes spread the disease from Ondonga in the southern floodplain to Oukwanyama in the north and next to southwestern Angola via the Kunene River fords. One observer claimed that only 1 % or 2 % of the African-owned cattle in the area survived. Southwestern Angola's cattle herds in Humbe and Gambos were estimated at half a million head before the rinderpest. Less than a decade later, Humbe contained a mere 50,000 head of cattle. The price of oxen more than quadrupled and all trade with ox wagons ceased. Carnivores preyed on humans for want of game. Rinderpest not only triggered a crisis in hunting, but also changed the nature of the cattle trade in the years immediately following the introduction of the disease because Africans refused to sell any breeding stock. Local diets and dress (leather had been the primary material used for clothing) also changed radically.[7]

---

[6] NAN, A233, J. Chapman collection, 1903–1916, pp. 61–62; and Peires, *The Dead Will Arise*, pp. 70–73.

[7] NAN, A233, J. Chapman collection, 1903–1916, pp. 45–47, 159–160, 167–169; "Ainda o Desastre do Humbe", and "A Peste Bovina em Angola", *Portugal em Africa* 5 (March 1898): 51, 128–

Cattle losses due to rinderpest in the Ovambo floodplain were probably much lower than the 90%-plus losses that European observers claimed. Migrating wildlife and cattle must have spread the disease across the Kunene River during the dry season, the only time when its fords were passable. During the dry season, however, cattle were dispersed over a large number of isolated cattle posts, which might have offered protection against rinderpest, since although it is highly contagious, transmission is dependent upon direct contact with diseased animals. In the Oukwanyama kingdom, for example, cattle losses due to rinderpest reportedly were in the hundreds or thousands rather than the tens of thousands which a mortality of 90% plus would have required. By all accounts, rinderpest affected critical animal resources with dire consequences. Cattle were the principal means to purchase war supplies (guns, lead, gunpowder, cartridges and horses) in an era of increased insecurity, especially since local sources of ivory and feathers had been depleted by the early 1900s. As competition for the remaining cattle increased, insecurity further mounted, triggering an arms race.[8] Despite attempts to make up for lost animal proteins by increasing crop production or fishing, insecurity negatively affected cultivation.[9] Depleted livestock herds made manure scarcer. Significantly, during the early 1900s, the inhabitants of the northern floodplain suffered from a series of famines.[10]

The steep decline of grazing and browsing wildlife in conjunction with the virtual absence of elephants due to hunting caused bush to encroach upon grasslands throughout the region, as noted in descriptions of the vegetation in the northern and central floodplain from the early 1900s. In 1881, the only dense forest described in Ombadja was alongside the Kunene River, where a path had to be cut to allow the passage of ox wagons. In 1907, the expanse of open plains in Ombadja

---

136; Marquardsen, *Angola*, pp. 99–101; CNDIH, Avulsos, Caixa 4121 "Humbe", 32–4.6 Humbe, Cicumscricão Civil: Agricultura 1891–1913, relatorios 1906; Schachtzabel, *Angola*, pp. 89–99; AVEM, RMG 2630 C/k 7, Wulfhorst, Referat: "Giebt es in unserer Ovambomission eine Frauenfrage?" Omupanda, Oct. 1903 (leather clothes); see also NAN, SWAA, Native Affairs Vol. 456, Secretary for South West Africa, "Native Cattle", Replies to Questionnaire by Dr. G. Schmid, GVO, n.p. [1932], to Secretary for External Affairs, Pretoria.

8 Kreike, *Re-creating Eden*, p. 46; Schachtzabel, *Angola*, p. 99.

9 Marquardsen, *Angola*, pp. 99–101.

10 Kreike, *Re-creating Eden*, chaps. 2–3.

was surrounded by dense and predominantly thorny species; the Portuguese invasion forces were forced to cut their way through the impregnable vegetation.[11]

The recovery of rinderpest-affected animal populations took two decades, and at least as much time passed before the region again was referred to as a "hunter's paradise".[12] And only another decade later, cattle raising in the southern floodplain once more was considered to be as important as crop cultivation.[13]

During the dry season, herdsmen took the cattle to the distant cattle posts beyond the Ovambo floodplain. After the first good rains, the herdsmen accompanied the cattle back to the villages. Concentrating the cattle in the villages had a negative side: contagious diseases could spread more easily. Lungsickness, for example, struck early during the rainy season, when open water was abundant throughout the floodplain and cattle were concentrated in the villages. Although lungsickness did not have the apocalyptic impact in Ovamboland that it had had in the Xhosa territory, it did become an increasingly severe problem. Or at least, the South African colonial administration perceived it as such, and consequently introduced measures to limit the export and movement of cattle. Prior to 1930, reports about lungsickness in Ovamboland were rare, but they became increasingly more common and more urgent.[14]

A third virgin soil cattle disease which appeared in Ovamboland in the second half of the 1940s, foot and mouth, jolted the colonial administration into taking radical actions and led to the consolidation of the Red Line, the veterinary barrier between Ovamboland and the rest of Namibia. The disease itself, however, did not cause any significant livestock losses at all: foot and mouth symptoms mirrored another innocent local disease that was well-known to Ovamboland's cattle holders.[15]

---

[11] NAN, NAO 104, C.L. Anderson to Hahn, 13/4/44 Johannesburg 1882, extract diary W.W Jordan; "Campanha do Cuamato", *Portugal em Africa* 14 (Sep. 1907): 165, 446–447. Cf. CNDIH, Avulsos, Caixa 3703 "Huila", Processo Missão de Estudos no Sul de Angola, 1914–15, Relatorio do Mez de Outubro [1914].

[12] Quadros Flores, *Recordações do Sul de Angola*, p. 200.

[13] NAN, NAO 18, Hahn, Notes on Ovamboland for the Administrator, Windhoek, 15 May 1924.

[14] NAN, NAO 18, Monthly Report Ovamboland, March 1925.

[15] See chapters 3 and 6.

Invading microbes and virgin soil epidemics

Several diseases also struck the human inhabitants of Ovamboland early in the colonial era in the 1920s. Three of them were triggered by invading microbes: influenza, measles and plague. In addition, TB and smallpox appeared. The TB incidence in Ovamboland increased as migrant laborers from the area were exposed to the disease and to unhealthy working and living conditions in southern Africa's mines.

Colonial reports are dismissive or vague about any virgin soil epidemics in 1920s and 1930s Ovamboland, and influenza and measles epidemics received little attention in colonial reports. In June of 1924, an influenza outbreak in central and southern Namibia virtually halted recruitment because, as an obviously annoyed official wrote, returning migrant laborers "brought back exaggerated reports of sickness on the diamond fields". The disease spread to Ovamboland, but according to the annual report for 1924, the death toll was not heavy. However, the very first colonial monthly report for Ovamboland, which covered December 1924, stated that influenza had afflicted "thousands" in the month of November alone, although no additional information was available about the disease for December. The following year, influenza caused "a great number" of deaths in Ovamboland. By October, the epidemic had killed at least two hundred people in the border district of Oukwanyama and fifty in the Ondonga district; it continued to rage in Ondonga in December 1925, where its victims included several headmen. Another influenza epidemic in October 1928 led to high but unspecified mortality, especially amongst children and the elderly. In 1931 and 1932, influenza struck again, in the latter year killing the prominent headman of Eenhana in eastern Oukwanyama. In 1935, influenza once more surfaced in the Oukwanyama and Uukwambi districts but again no details were provided.[16]

A 1927 outbreak of smallpox in Angola led to a vaccination campaign in Ovamboland that progressed smoothly in Ondonga and the northwestern districts and prevented the disease from becoming an epidemic south of the border.

[16] NAN, NAO 18–19, Annual Reports Ovamboland 1924 and 1925, and Monthly Reports Ovamboland, Dec. 1924, Sep.-Dec. 1925, Oct. 1928, May – June 1931, Sep. 1932 and June – July 1935.

In 1942, the Oukwanyama district experienced an outbreak of 'Kaffir-pox', and the response of the administration was to vaccinate people in the affected areas, although with a delay of three years, possibly related to the war conditions. Almost 54,000 people were vaccinated against smallpox in 1945.[17]

Two epidemics of measles, in 1929 and in 1938, are only mentioned in passing in the colonial records, although local memories associate an outbreak of measles on the Angolan side with heavy mortality amongst recently initiated girls. The annual report for Ovamboland for 1929 mentions no deaths, stating "[no epidemics to report] with the exception of a few cases of measles". But the small colonial staff of Ovamboland had not only been distracted by routine tasks, but also had been overwhelmed by a famine and the introduction of a tax system. In addition, its medical staff was shorthanded because one of the administration's two medical officers had been temporarily re-assigned to the Caprivi region to attend to a smallpox outbreak. As a result, the impact of the 1929 measles epidemic may have been underreported. In July – August 1938, the Anglican mission reported that an outbreak of measles at its Holy Cross Mission had spread rapidly throughout the Oukwanyama district, causing several deaths. The outbreak was serious enough for the administration to halt migrant labor recruitment for a few weeks. No female initiation ceremonies were reported in that year. Women's initiation ceremonies took place in Ovamboland in October 1928, despite a threatening drought: "In spite of the depressing conditions wedding feasts are being celebrated with much beer and noise, this being the season of marrying and taking in marriage".[18] It is possible that the measles epidemic that is known as the killer of 'brides' (as recently initiated girls were referred to) took place in 1929.[19]

A third disease introduced alongside colonialism was cause for concern to the administration. Tuberculosis patients had been treated within the (government-subsidized) mission hospitals in Ovamboland since at least 1927. Questioned

[17] NAN, NAO 18, Monthly Reports Ovamboland, Aug.-Sep. 1927; A450, 7, Annual Report Ovamboland 1945; NAO 60, Quarterly Report Ovamboland, July-Sep. 1952.

[18] NAN, NAO 18, Monthly Reports Ovamboland, Oct. 1928.

[19] NAN, NAO 18, Monthly Report Ovamboland, Oct. 1928 and Ovamboland Annual Report 1929; NAO 20, Monthly Reports Ovamboland, July – Aug. 1938. Maria Weyulu, interview by author, Eenhana Refugee Camp (Namibia), 16 July 1993.

about health issues in Ovamboland by a commission of inquiry, the District Surgeon for Ovamboland stated that he had treated 79 cases of TB in Ovamboland in 1934 and that 17 patients had died. He added that "according to histories given by patients, it appears as if many of them contract the disease while at work in the South [i.e., as migrant laborers on the farms and in the mines of Namibia]".[20] Tuberculosis remained endemic in Ovamboland: in 1951, the hospitals in Ovamboland treated 257 cases, and 35 patients died.[21]

A fourth invasive microbe that made its way into Ovamboland during the 1930s, however, jolted the South African colonial administration into a frenzy of action. In January 1932, the Finnish Mission Society, which ran several hospitals and clinics in Ovamboland, reported the outbreak of what appeared to be plague. A high-ranking medical official from the South African Health Department was immediately flown in and hastened to Ovamboland to investigate the cases with the District Surgeon. The two doctors collected blood smears, sending the samples to a lab for analysis while the death toll rose to twelve. The blood samples confirmed the worst fears: it was bubonic plague. The plague continued to spread as the administration brought in plague serum and vaccines. The administration also dispatched a special rodent inspector to Ovamboland to combat the disease's main vector. The inspector received his own motor vehicle, a rare luxury especially in this era of economic depression, and indicative of the extent of the administration's concern about the outbreak. The Portuguese authorities stationed two of their precious medical officers on the border to monitor the disease in Ovamboland and the officers remained in close touch with their South African counterparts. The Portuguese also closed the Angolan border "to all travelers and placed a chain of military outposts along the line" that served as a base for regular border patrols. Sealing the border triggered a series of incidents. In one instance, Ovamboland residents returning to Namibia from Angola with their cattle disarmed a Portuguese indigenous soldier guarding the border. Fearful of an escalation of the violence, the Portuguese withdrew their patrols soon thereafter.

---

20 NAN, A450, 12, SWA Commission: Minutes of Evidence (1935), vol. 12, Session at Ukualuthi, 13 Aug. 1935, evidence by Dr. Michiel van Niekerk (District Surgeon Ovamboland and Kaokoveld); and NAO 18, Native Affairs Ovamboland Annual Report 1927.

21 NAN, NAO 65, Annual Health Report Ovamboland and Kaokoveld 1951.

The number of plague cases declined rapidly after February 1932 but increased again by the end of the year. By mid-1933, the total cumulative death toll of bubonic plague had risen to 61, representing one of every five people diagnosed with the disease. Ondonga was the most severely affected district, with more than half of the confirmed cases and half of the fatalities.[22] Plague returned in 1934, but it was no longer confined to remote villages, causing resident colonial officials to fear for their own lives: "[Flea vectors] are now a constant source of danger to the Europeans".[23] Migrant laborers recruited in Ovamboland during 1934, 1935 and 1936 received "prophylactic injections" against plague before they were allowed to leave the reserve to engage in wage labor in the 'white' regions of colonial Namibia.[24] In 1935, plague struck again, killing 27 of 175 infected people.[25] Cases of plague occurred again in 1937 and a "severe outbreak" of what is identified as "pneumonic plague" was reported in 1943, but without details about any fatalities.[26] It is unclear whether the 1943 outbreak was indeed the less deadly pneumonic plague, or whether it was reported in error, because the lab analysis of the blood smears collected during the initial outbreak in 1932 had identified bubonic plague. Plague and its rodent carriers continued to worry the Ovamboland administration. In 1948, the colonial staff still included a Rodent Inspector and in the mid-1950s, officials issued what appears to have been a routine request to Ovamboland's Tribal Councils to monitor rodent populations closely and to report any renewed outbreaks of plague.[27]

Although colonial reports argued that the epidemic was due to the local practice of consuming rodents, which they described as disgusting and unhygienic, the actual source was food aid shipments of maize from South Africa with infected

---

[22] NAO 19, Annual Report Ovamboland 1932 and Monthly Reports Ovamboland, Jan., Feb., April, July – Dec. 1932, June – Oct. 1933.

[23] NAN, NAO 19, Monthly Report Ovamboland, March 1934.

[24] NAN, NAO 19–20, Monthly Reports Ovamboland, Sep.-Oct. 1931, Jan. – Feb. 1932, March – Aug. 1932, June and Oct. 1933, March and Oct. 1934, March and June – July 1935, Jan. – March and June – July 1936, and Quarterly Report Ovamboland, Jan. – June 1943; NAO 36, District Surgeon Ondangwa, Annual Report 1933.

[25] NAN, NAO 19, Monthly Reports Ovamboland, Jan. – March and June – July 1935.

[26] NAN, NAO 36, Annual Health Report Ovamboland 1937 [or 1936?], and NAO 21, Quarterly Report Ovamboland, Jan. – March and April – June 1943.

[27] NAN, NAO 66, Rodent Inspector Ovamboland to District Surgeon Ovamboland, Ondangwa, 18 June 1948; and Tribal Secretary Ondonga to NCO, Okaloko, 20 July 1954.

rats during the early 1930s Famine of the Dams.[28] The plague became endemic in the region because Ovamboland's abundant indigenous rodent population rapidly became infected with the disease: fleas carried by the rodents were the vector that transmitted the disease to humans. More recent outbreaks of plague in the Ovambo floodplain occurred in 1992 and 1993.

A plague of donkeys: Fierce invading equines

Horses played an important role in the story of the military and biological conquest of the Americas. Hernando Cortés had approximately two dozen war horses that inspired great terror amongst his Aztec opponents. When adopted by the North American Indians, horses revolutionized the local economy and society, initiating the era of the dominance of the Plains Indians.[29] But the horse, ridden by cowboys and the U.S. Cavalry, was also powerfully helpful in the conquest of the Indians and the establishment of a Neo-Europe.

Horses were also an important tool of empire and biological conquest in Africa. The great savanna empires of medieval West Africa—including Ghana, Mali and Songhay—relied on cavalry. Cavalry played a role in the destruction of Songhay by Moroccan invaders. Horses were imported from North Africa and Europe, commanding high prices, and they also were bred locally. Cavalry remained important in warfare well into the nineteenth century, facilitating (slave) raids and conquest. In his resistance to the French advance in the latter nineteenth century, the West African resistance leader Samory relied heavily on cavalry, and cavalry also played an important role in the French conquest of the savannas of West Africa.[30]

The Portuguese not only supplied West African rulers and merchants with horses, but also introduced horses in southern Africa. The numbers of horses remained small in both modern Mozambique and Angola and their impact was more akin to a Cortés-style weapon of terror.[31] In South Africa, horses (and guns)

[28] South African National Archives Pretoria, NTS Ovamboland, PM 1/2/176, PM49/91. These materials have since been transferred to the National Archives of Namibia in Windhoek.

[29] See Isenberg, *The Destruction of the Bison*.

[30] Law, *The Horse in West African History*, and Goody, *Technology, Tradition, and the State in Africa*. On Samory, see Person, *Une révolution Dyula*.

[31] See, for example, Amaral Ferreira, "Transforming Atlantic Slaving".

afforded the Boers, the Griqua, the Bastards and the Oorlam great mobility and power.[32] In fact, the Boers probably only survived in the interior of South Africa because of the combination of horses and firearms.

In the 1860s and 1870s, the Ovambo floodplain and the wider region were subjected to horse-mounted and gun-armed raiders from central Namibia. Terrorized and impressed, in the following years the floodplain's elite invested heavily in purchasing guns and horses, even though both commanded very high prices. Before the 1897 rinderpest, a horse's value ranged from 70 to 190 head of cattle. Both high prices and the prevalence of horsesickness limited their numbers to several hundred at the most from 1850 to 1930.[33] The Portuguese and South African invading armies made heavy use of cavalry in the floodplain but they limited their operations to the dry season.[34]

Horsesickness was mentioned frequently in regular reports from the colonial administration in Ovamboland, from their inception in 1924 onward. The reports record that horsesickness occurred during the rainy season and that when there was an abundance of standing water, losses were especially severe. During the rainy season of 1924–1925, the outbreak of horsesickness was first reported in December 1924, and losses steadily mounted until into the month of May. By the end of April, when standing water still abounded, losses to horsesickness reached 93 animals. By February 1927, horsesickness had returned, killing "several" horses. Because of their scarcity, horses continued to be highly coveted, but because they no longer were valued in warfare, the cost was a fraction of the pre-1890s rinderpest price. In 1932, a horse cost 6–7 head of cattle, as against the 70–190 head of cattle before the advent of rinderpest.[35] In 1951, the headmen of the Oukwanyama district purchased 285 doses of horsesickness vaccine and it was reported that "many of the horses have been inoculated".[36]

---

[32] See Ross, *Adam Kok's Griquas*; Lau, *Southern and Central Namibia in Jonker Afrikaner's Time*; Swart, "'Horses! Give Me More Horses!'".

[33] Kreike, *Re-creating Eden*, p. 38, table 2.1. On horsesickness, see NAN, NAO 18, Monthly Reports Ovamboland, Jan. – Feb. and April 1925, Feb. 1927. Cf. NAO 104, Anderson to Hahn, Jordan diary.

[34] Kreike, *Re-Creating Eden*, pp. 35–80.

[35] NAN, NAO 18, Monthly Reports Ovamboland, Dec. 1924, Jan. – May 1925, Feb. 1927, and NAO 16, statement of Tiolene, appendix to O/C Oshikango to NCO, Oshikango, 22 June 1932.

[36] NAN, NAO 18, Monthly Reports Ovamboland, Dec. 1924, Jan. – May 1925, Feb. 1927, and NAO 16, statement of Tiolene, appendix to O/C Oshikango to NCO, Oshikango, 22 June 1932; NAO 60, Quarterly Report Ovamboland, Oct. – Dec. 1951.

Mules and donkeys were also introduced in Ovamboland early during the colonial period, and they too remained few in numbers until the 1930s.[37] Like horses, mules were vulnerable to horsesickness, but, the extent to which the disease affected donkeys is unclear.[38] After World War II, however, the number of donkeys quickly expanded and colonial observers began to describe donkeys as a veritable plague. In the South African Bophuthatswana homeland, donkeys were culled in the 1980s 'donkey massacre'.[39] Today, donkeys are identified as one of the most dangerous and prevalent invading species in Ovamboland, and a severe environmental threat.

The history of donkeys defies conventional narratives of invasive species, including that of Latin America's plague of sheep.[40] Whereas the number of sheep in Mexico quickly increased following their introduction, donkeys in Ovamboland initially did not fare much better than their equine cousins. Until the late 1930s, the floodplain environment did not seem to favor donkeys. A September 1937 colonial report noted that donkeys

> do not seem to thrive, at all well, in Ovamboland. In former years it was not so noticeable but whether there is a difference in the grazing or watering conditions, is not known, but the fact remains that all donkeys, as soon as the grass discolours and becomes dry, rapidly fall off in condition. They all appear to be suffering from intestinal trouble. This according to the Veterinary Officers with whom I have discussed this matter must be due to bad water. It may be that the successive flood seasons, experienced since 1934, brought with them some parasite, from the North, which is injurious to donkeys.[41]

Donkeys and mules were used as transport animals. Donkey carts were much more versatile than ox wagons because the prevalence of lungsickness in Ovamboland meant that oxen could not be used to transport goods between Ovamboland and the rest of Namibia. In the 1930s, the Anglican mission at Odibo near Oshikango used donkey wagons to haul supplies into Ovamboland from

---

[37] NAN, NAO 18, Monthly Reports Ovamboland, Feb. 1925, Aug. 1926 and Aug. 1927.
[38] NAN, NAO 18, Monthly Reports Ovamboland, Dec. 1924, Jan. – Feb. 1925.
[39] Jacobs, *Environment, Power and Injustice*.
[40] Melville, *A Plague of Sheep*.
[41] NAN, NAO 20, Monthly Report Ovamboland, Sep. 1938.

the railhead at Outjo. The colonial administration and the local population used donkeys to carry supplies and water.[42]

Increased opportunities for paid labor during World War II—including military service—allowed more people to invest in donkeys and carts. In 1942 and 1943, increased military service wage returns brought over 8,000 pounds sterling per month into Ovamboland at the same time that the availability of imported manufactures in the local stores declined. Many returning migrant laborers bought donkeys in the white farming regions south of Ovamboland "at ridiculously low cost" to carry food and other products home. In 1943, the Senior Headmen of Ovamboland allegedly asked the Native Commissioner to limit the import of donkeys because "[t]hey feel that they may easily become a menace to grazing in the same way as has happened in certain areas in SWA". A 1944 census revealed that the Oukwanyama district alone had 3,410 donkeys while the total number of donkeys in Ovamboland in 1943 was estimated at over 8,000 animals. Native Commissioner Hahn claimed to have met Africans who brought in 500 donkeys at the time, and he discussed the matter with the headmen:

> All have expressed concern at the position, particularly on account of the shortage of grazing for cattle and other stock. In the vast majority of cases the natives acquire donkeys for the purpose of carrying their goods from the railhead and on reaching Ovamboland they are allowed to run wild and are seldom used again.

Hahn wanted to prohibit the passage of donkeys via Ovamboland to Angola. In March 1945, with the agreement of at least the Oukwanyama Council of Headmen, a ban on the import of donkeys into Ovamboland came into effect. Exceptions would only be made for draft animals that had been purchased in the Police Zone.[43] In 1945–1946, the number of donkey carts in Ovamboland was estimated at 200 and the number of donkeys was estimated at 6,000.[44] The 1950 census report

[42] MacDonald Diary, 1932–1944 (Private Collection Nancy MacDonald, courtesy Nancy MacDonald), Odibo; NAN, NAO 20, Monthly Reports Ovamboland, Aug. 1937, Aug.-Sep. 1939.

[43] NAN, NAO 43, NCO to Graig, Ondangua, 22 June 1943; NAO 15, ANC to NCO, Oshikango, 14 Nov. 1944, NCO to CNC, Ondangwa, 22 Nov. 1944 and NCO to Secretary SWANLA, 3 Jan. 1945; A450, 7, Annual Reports Ovamboland 1942 and 1943. On the use of donkey carts, see BAC 44 f. 1/15/4/17, Minutes Annual Meeting Ondonga, 1 Dec. 1958 and BAC 45, Minutes Tribal Meetings Oukwanyama, 9–29 May 1958.

[44] NAN, NAO 103, Census of Agriculture, Ovamboland 1945–1946.

only listed 50 donkey carts, but an actual count in 1952 produced a total of 576 single-axle donkey carts, including 136 in Oukwanyama, 100 in Ondonga, 76 in Ombalantu, 152 in Uukwambi, 108 in Ongandjera and 4 in Onkolonkathi.[45]

During the 1950s, donkeys continued to be used as pack animals, for example, by hunters to carry meat and by female petty grain traders.[46] Using the donkey as a plowing animal seems to have been a fairly recent 1950s innovation.[47] Low prices in the Police Zone and the general usefulness of donkeys (despite colonial perceptions to the contrary) encouraged migrant laborers to use them to transport their goods back to the Ovambo floodplain. The number of donkeys in Ovamboland consequently increased sharply in the 1950s and 1960s. In 1966, the Namutoni Farmers Association of the Tsumeb district (an organization of white farmers) complained that migrant laborers who traveled through their district on their way home were accompanied by donkeys loaded up with furniture and other goods. The Police Commander at Oshivelo, the main entry point into Ovamboland for returning migrant laborers, estimated that since the new Oshivelo road had opened six months ago, 1,500 donkeys and 1,500 horses had passed his post.[48] An increase in the donkey population was also facilitated by what appears to have been a decreased vulnerability of the animals to the floodplain disease environment. By the late 1950s, they reportedly rarely succumbed to disease, although horses continued to suffer from a high death rate.[49]

Overgrazing came to be seen as an increasingly serious threat to north-central Namibia's environment during the 1940s. Such radical measures as livestock culling, which had been attempted elsewhere in southern Africa (and Kenya), however, were not introduced in Ovamboland. In 1954, the Administrator of

---

45 NAN, NAO 103, Census of Agriculture Ovamboland, 1949–1950, and ANC to NCO, Oshikango, 30 Dec. 1952; Chief Kambonde to NCO, Okaroko, 18 Dec. 1953; Council of Headmen of Ombalantu to NCO, 25 July 1952; Council of Headman of Ukuambi to NCO, Ukuambi, 16 July 1952; Chief Ushona Shimi to NCO, Okakua, 7 July 1952; Ikasha Nkandi and Ashimbanga Mupole to NCO, Onkolonkathi, 26 June 1952.

46 NAN, NAO 90, Statements by Johannes Shekudja before NCO, Ondangwa, 18 and 19 March 1952; NAO 62, ANC to NCO, [Oshikango], 22 Oct. 1952, and Pastor Risto Ushona to NCO, Okaku, 8 Sep. 1952.

47 NAN, NAO 100, Ruusa Amtenya vs. Nikodemus Amtenya, statements by Ruusa Amtenya and Nikodemus Amtenya, Ondangwa, 17 Sep. 1954.

48 NAN, AHE (BAC) 332, Report of Meeting of Namutoni Farmers Association, Tsumeb District, 25 June 1966.

49 NAN, BAC 133, Agricultural Report Ovamboland 1956–1957.

colonial Namibia visited Ovamboland and in public meetings criticized the environmental abuses of Ovambo farmers. He singled out donkeys explicitly:

> I also want to warn you against another thing. You should not keep donkeys. The donkey does not only eat the grass but he also uses his hoofs to scratch out the roots. We also had many donkeys initially, but we realized how bad they were and we then sent them to the factories / plants to be slaughtered.[50]

Donkeys remained a target of the colonial administration throughout the 1950s and 1960s. In 1958, the Chief Native Commissioner compared donkeys (and horses) with locusts and imposed a temporary ban on importing donkeys that lasted throughout the dry season.[51]

The warnings and measures, however, had little or no effect. Donkey numbers rose to 25,000 in 1956–1957, and 31,382 in 1959. In the 1960s, the statistics suggest a decline in the number of donkeys to 25,500 animals in 1966 and 18,000 in 1967. The number was still 18,000 in 1968, although another report from 1968 set the number of donkeys, horses and mules at 35,000. The late 1960s show an increase to 35,000 in 1969, and 55,000 in 1970, followed by a decrease in the early 1970s (50,611 in 1971, 52,540 in 1972, 56,236 in 1973), increases in 1974 (to 60,958) and 1975–1976 (to 86,227) and a dip in 1981 (to 81,000). Although the figures—all estimates—suggest years of rapid increase followed by decreases between individual years, the overall multi-year trend is a rapid increase from 6,000 in 1945–1946 to 81,000 in 1981, a growth by a factor of 13.[52]

---

[50] NAN, NAO 64, Minutes Ukwanyama Tribal Meeting attended by the Administrator [12 July 1954].

[51] NAN, BAC 44, Minutes Annual Meeting Ondonga, 1 Dec. 1958. Cf. BAC 45, Minutes Tribal Meetings Oukwanyama, 9–29 May 1958. In 1961 the import of donkeys was again prohibited, BAC 131, Quarterly Report of Meeting held at Outanga, Oundonga [Ondonga], 27 Dec. 1961.

[52] NAN, BAC 133, Agricultural Report Ovamboland 1956–1957; WAT 1, "Equines, goats, pigs, and poultry in Northern Native Territories outside Police Zone, 30 June 1959"; Ovamboland; AHE (BAC) 1 / 352 [1 / 357], Annual Report Agriculture Oukwanyama for 1964 and Annual Reports Agriculture Ovamboland for 1966 and 1968; OVA 49, Agricultural Statistics 1967, appendix to Chief Director Department of Economic Affairs to Director Agriculture, Ondangwa, 25 March 1967; OVA 9, Chief Agricultural Official to Director Agriculture, Ondangwa, 25 June 1969, "Regarding Questionnaire", and OVE 10, table 6.1.3. The 1970, 1971 and 1975–1976 figures included horses and mules which from 1972–1974 numbered 1,200–1,500, see OVA 40, questionnaire appended to Secretary Economic Affairs to Secretary of Agriculture, Ondangwa, 10 Nov. 1973 and OVE 10, table 6.1.3, Veegetalle in Owambo, 1968–1972; OVA 6, Annual Report Veterinary Service Owambo 1975–1976; OVA 9, Director Agriculture and Forestry to Director-General Department of Cooperation and Development Pretoria, [Ondangwa], 5 May 1981.

Not only did donkeys continue to serve as versatile transport animals, but during the 1950s and 1960s, north-central Namibian farmers discovered the donkey's potential as a cheap plowing animal in an era when plow technology was rapidly spreading. Colonial officials continued to rail against donkeys and again threatened to prohibit their import at meetings in Oukwanyama and Ondonga in 1958, but at both of the meetings, people in the audience stressed the importance of donkeys as plow animals. At the meeting in the Ondonga district, Johannes Kuandambi said: "it hurts to hear these words about donkies because they are our assistents that we use to plough and they transport our goods and the meat of cattle that is dying at the moment". At the meeting in the Oukwanyama district, the Native Commissioner retorted that oxen should be used to plow and "if any man has not enough oxen of his own he should pool his with those of his neighbour or neighbours. Such communal ploughing works very well in the Union [of South Africa]".[53] In 1968, the Director of Agriculture worked out an agreement with the Tribal Councils of Ovamboland whereby no further donkeys could be imported without a purchase and import permit issued by the councils. But migrant laborers continued to illegally purchase donkeys and smuggle them into Ovamboland.[54]

In practice, the import of donkeys and horses was only effectively interfered with during drought years.[55] In May 1971, the state veterinarian based at Otavi to the south of Ovamboland complained that every year Africans bought "uncountable" numbers of horses and donkeys in South West Africa—the latter from as far south as Mariental—and drove them in herds on sometimes months-long drives to Ovamboland.[56] Extension officers castrated small numbers of donkeys in the early 1970s, but the practice appears to have been confined mainly to animals owned by headmen. For example, one of the two extension officers in

[53] NAN, BAC 44, Minutes Annual Meeting Ondonga, 1 Dec. 1958, and BAC 45, Minutes Tribal Meetings Oukwanyama, 9–29 May 1958. See also BAC 133, Agricultural Report Ovamboland 1956–1957, and O/C Native Affairs Kaokoveld to CNC, Ohopoho, 4 Oct. 1950; Census of Agriculture, 1949–1950, Ovamboland and NAO 61, [ANC] to NCO, n.p., 9 April 1949.

[54] NAN, AHE (BAC) 332, Director Agriculture to Chief Bantu Commissioner, Windhoek, 20 Aug. 1968.

[55] NAN, AHE (BAC) 332, Director Agriculture to Chief Bantu Commissioner, Windhoek, 20 Aug. 1968 and AHE (BAC) 332, Secretary Department of Agriculture to Chief Bantu Commissioner, Ondangwa, 25 March 1974 and Chief Bantu Commissioner to Director Agriculture Ondangwa, Windhoek, 4 March 1974.

[56] NAN, OVA 56, State Veterinarian to Chief Director Owambo Government, Otavi, 5 May 1971.

Oukwanyama district castrated 41 donkeys in 1973 and 56 in the following year, while in May – June 1976, another extension officer castrated 151 donkeys. But the district counted over 11,000 donkeys in 1971 so the impact of the extension officers' efforts was minimal.[57] Donkeys, mules and horses did suffer from intestinal parasites and in early 1970s Oukwanyama, an extension officer reported that people had requested the distribution of such deworming medicines as Askaritox and Equizole.[58]

In 1974, in western Oukwanyama three village headmen had some of their livestock castrated: one of them brought in 7 bulls and one donkey, a second, 5 bulls, 2 billy goats and 9 donkeys, and a third, 7 bulls and 2 donkeys.[59] In 1975, only one of a sample of four wealthy livestock owners in the far eastern Oukwanyama had a pair of donkeys.[60] Out of a 1993 sample of 101 households surveyed by OMITI, 10 owned one donkey, 19 owned 2 donkeys, 14 owned 3, 11 owned 4, 7 owned 5, 6 owned 6 and one each owned respectively, 17, 28 and 60 donkeys. Their owners mostly kept the donkeys near their farms; only 4 owners kept a total of 30 donkeys away from their homes, while 86 other owners kept their total of 176 donkeys close to home.[61] Of the 69 households that answered a question in the survey about herding, 47 claimed to have herded their donkeys during the rainy season, but donkey owners did not herd their animals during the dry season. Unherded animals were sometimes tied (2 out of 70 mentioned this), which probably refers to the not uncommon practice of hobbling—tying the front legs of a donkey together with a short rope to prevent it from wandering too far. The household's children were usually the herders, or alternatively the head of the household or his wife watched over their donkeys.[62]

---

57 NAN, OVA 61, Monthly Reports Supervisor Agriculture: Andreus Ndeitwa (Ohangwena, Oukwanyama), 1971–1976; OVA 59, Monthly Reports 1974 and 1976 [Agriculture], Leonard Haihambo, Oukwanyama. See also OVA 61, Monthly Reports Agricultural Officer: Moses Nandjebo [Ohangwena] (Oukwanyma), 1973–1975.

58 NAN, OVA 61, Monthly Report for August Supervisor Agriculture: Andreus Ndeitwa (Ohangwena, Oukwanyama), 1971–1976.

59 NAN, OVA 61, Monthly Reports 1974 Supervisor Agriculture: Andreus Ndeitwa (Ohangwena, Oukwanyama).

60 NAN, OVA 61, Monthly Report Feb. 1975, Agricultural Officer: Moses Nandjebo [Ohangwena] (Oukwanyma).

61 OMITI survey, 2.51.

62 OMITI survey, 2.5.2.0–1 and 2.5.3. For hobbling donkeys, personal observations, 1991–1993.

Almost all households (101) in a sample of 115 from the OMITI survey used donkeys for plowing; almost one-half also used the animals for carrying water (54) and pulling carts (51), and one third (36) used them to carry firewood. Thirteen respondents mentioned other transport functions, and only 10 respondents mentioned breeding and meat production.[63] The 1979–1980 agriculture report stated that the tribal authorities of Ovamboland owned one plow, that private farmers owned a total of 100 plows, and that 20,000 ha (of a total of 190,000 ha) were plowed.[64] The figures seem to be a gross underestimation of the number of animal-drawn plows in Ovamboland, which may partly be a reflection of the increased use of donkeys as opposed to oxen as draft animals: donkey plows may have been simply not included in the statistics.

The principal feed of donkeys consisted of grass (mentioned by 100 % of a 113-household sample in the survey), woody vegetation (75 %) and roots (19 %).[65] Donkey users were fully aware of the environmental costs of using the animals: 98 out of 120 households in the survey emphasized that donkeys fed differently than cattle; 64 observed that donkeys ate more than cattle; and 14 respondents pointed out that donkeys simply ate too much. Respondents also explained that not only did donkeys eat larger quantities of grass than cattle, but unlike cattle, they were not at all selective in what they fed on and they ate continuously, even at night. Respondents also observed that donkeys cropped grass much closer than cattle, and that they dug out roots and scraped bark off trees, and in 12 cases, the surveyors understood from the respondents that the latter believed that donkeys caused overgrazing.[66]

The large majority (127) of a 158-household OMITI sample believed that the number of donkeys had increased since their youth.[67] Most respondents attributed the increase to the donkeys' many uses, especially as plowing animals (plowing was mentioned by 42 out of 113 respondents). In addition, in contrast to cattle, donkeys were not killed for meat (mentioned by 19 respondents) which in turn facilitated their increase, and, moreover, they were imported in large

---

63 OMITI survey, 2.5.4.
64 NAN, OVA 6, Department of Agriculture and Forestry Owambo, Annual Report 1979–1980.
65 OMITI survey, 2.5.5.
66 OMITI survey, 2.5.6.
67 OMITI survey, 2.5.7.

numbers from south of Ovamboland (mentioned by 31 respondents), where they continued to be cheap. One respondent even said that "donkeys produce more than cattle", i.e., donkeys are more productive than cattle, which was a very strong statement in a society where cattle have an enormous economic and symbolic value.[68]

Fierce indigenous creatures on the rampage

While many of Ovamboland's inhabitants embraced the donkey even as it was condemned as an invasive plague by colonial officials, the reverse was true for African indigenous 'royal game'. Colonial officials actively sought to protect elephants and lions, whereas most of their colonial subjects considered these animals dangerous predators, imperiling their lives, livestock and crops. But the South African colonial administration—despised for its apartheid policies but at times celebrated for its 'progressive' conservation record—considered other indigenous predators to be vermin that should be eradicated, including the cheetah and the wild dog. Today, both species are amongst the rarest predators alive, and the object of extensive conservation projects. Moreover, from World War II onwards the administration also characterized Ovamboland's indigenous domestic goats as despicable environmental vermin, although, like the donkey, the local population considered goats to be a critical resource. Thus, interpretations of which animals were a blessing or a curse differed radically between the colonial authorities on the one hand and the local population on the other.

Three factors brought humans, their domestic animals and their villages in closer contact with wildlife from the 1920s to the 1960s. First, humans encroached on wildlife habitat in and around the middle and southern floodplain. In the 1910s, 1920s and 1930s, thousands of refugees poured into the middle floodplain, an area that had been a wildlife paradise in the 1870s and 1880s. Secondly, as security improved in Ovamboland from the 1920s onward, inhabitants from the core areas of the precolonial polities fanned out into the borderlands that had

[68] OMITI survey, 2.5.8.

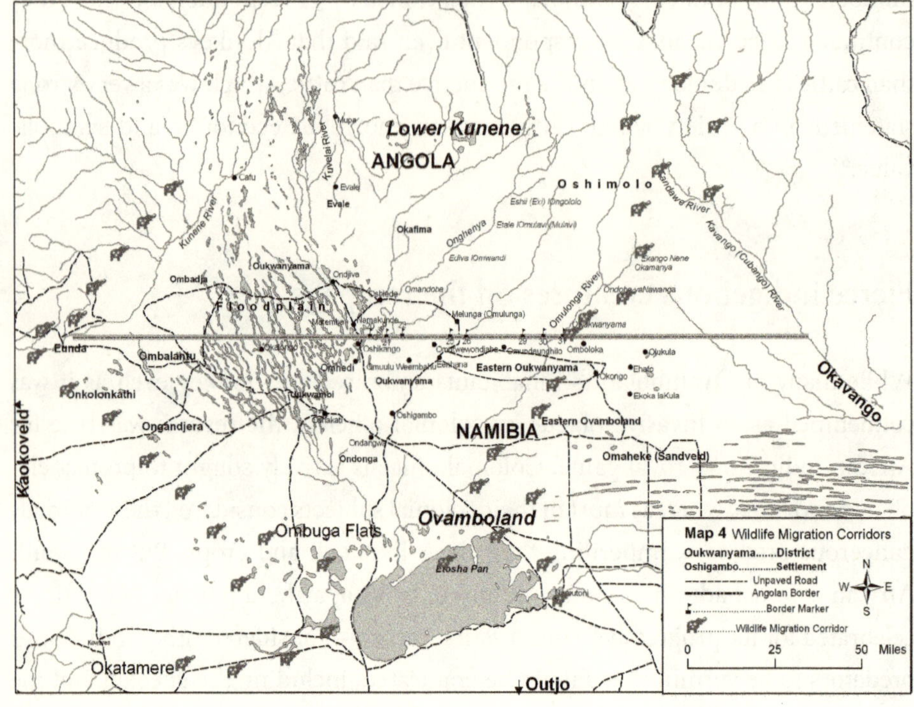

separated the polities, and to the edges of the floodplain and beyond. The influx of refugees and migrants from the Angolan side of the border and the rebuilding of cattle herds also led to the creation of more cattle posts further away from the floodplain.[69]

Thirdly, the border region's game populations, especially predators and elephants, recovered in the early colonial era because the local population was disarmed and colonial conservation measures were introduced. The sequence complicates progressive linear Nature-to-Culture narratives because the turn-of-the-century wildlife populations probably were at a low due to heavy hunting and the rinderpest, and they increased in the early colonial era. As refugees and migrants and their cattle permeated beyond the densely settled heartlands that had marked the precolonial Ovambo polities, predators, elephants and other wildlife simulta-

[69] See chapter 2 and Kreike, *Re-creating Eden*.

neously repopulated the very same areas. The competition between humans and their livestock on the one hand and wildlife on the other was especially keen whenever and wherever the settlement frontier intersected with wildlife migration routes.

Colonial conservation of 'royal game' intensified the conflict: colonial officials insisted in preserving lions and elephants at almost all costs, even if they caused tragic losses of lives or undermined livelihoods. Only in exceptional cases did colonial officials permit the killing of selected 'problem' animals, although the predators caused severe livestock losses, and elephant raids on fields, fruit trees and food stores on the edges of the expanding settlement frontier increased.[70]

Most of the big game in the Angolan-Namibian border region migrated between or around three major nuclei: Etosha Pan and the Ombuga Flats in the far southern floodplain, the Kunene River valley west of the floodplain, and the Oshimolo swamps and the Kavango / Cubango River valley east of the floodplain. The creation in 1928 of what was to become the Etosha Game Park (Game Reserve No. 5) made any game hunting within its vaguely defined boundaries illegal, although 'poaching' remained a problem in Etosha until the 1950s.[71]

The western game corridor connected Etosha Pan to the Kunene River which the animals crossed through numerous fords. Wildebeest, zebra, springbok, gemsbok, duiker and hartebeest were the most abundant animals during the 1930s and 1940s. They calved with the onset of the rainy season in their dry season haunts around Etosha Pan. Early in the rainy season, they moved to the open grass savanna of the Ombuga Flats to the north of Etosha Pan. When the Ombuga Flats with their poorly drained soils were too soggy, the large herds dispersed over a wide area to the west and northwest of Etosha, close to the inhabited areas of western Ovamboland, and as far west as the Kunene River valley. If standing water in western Ovamboland hampered movement and grazing, animals also fanned out to the sandy (and better-drained) Okamatere, directly west of Etosha Pan, and the Sandveld to the east. Towards the end of the rainy season, around April,

[70] Timotheus Nakale, interview by author, Ekoka laKula, 21 Feb. 1993; NAN, NAO 19–21, Monthly and Quarterly Reports Ovamboland, 1931, 1935, 1939–1941, 1946; NAO 60–61, Quarterly Reports Ovamboland, 1947–1949, 1952–1954.

[71] NAN, NAO 20, Monthly Reports Ovamboland, Jan. – Feb. 1937, and NAO 60, Quarterly Report Ovamboland, April-Sep. 1947.

wildebeest and zebra once more concentrated on the Ombuga Flats, followed by springbok, gemsbok and ostrich. The timing of this movement depended entirely on the availability of standing water in the Flats. In 1943, for example, surface water remained so abundant throughout the southern floodplain that in August the herds had not even begun to concentrate on the Flats. By contrast, early in 1946, rainfall was scarce and the big game herds already commenced returning to Etosha Pan. In June – July, the herds usually left the Flats for the permanent water holes south of Etosha Pan between Okakueyo and Namutoni. In general, gemsbok, hartebeest and kudu traveled along the southern edge of the western game corridor and along the edge of the Ombuga Flats because they preferred to browse on bush vegetation that was more easily found away from the Flats. Impala ventured from their dry season habitat in the Kunene River valley as far southeast as Etosha Pan during a good rainy season. Elephant, giraffe and eland shared the western corridor as they migrated between Kaokoland, the Kunene River and Etosha Pan.[72]

Elephants used the northern side of the western game corridor, closest to the inhabited areas of the southern floodplain. Traveling in larger and smaller groups, they skirted and, if forage and water was scarce, invaded the villages on the edge of western Ovamboland. As elephant populations rebounded and human settlement pushed into the fringes of the southern floodplain, competition over water and forages mounted. Mopane bush—a favorite of elephants and a dry season forage of cattle—was abundant in these areas. Elephant populations on the western corridor also increased because of hunting pressure across the Kunene River in southwestern Angola. During 1939 and 1940, elephants extended their range north of Etosha Pan.[73] During the 1940s and 1950s, western Ovamboland (Onkolonkathi, Eunda, Uukwaluthi and Ongandjera) and Ondonga, adjacent to the western game corridor from Etosha to the Kunene

[72] NAN, NAO 18, 20–21, Monthly and Quarterly Reports Ovamboland, July 1929, 1931, 1937, 1939–1946.

[73] Boer hunters from Angola smuggled ivory into Ovamboland via African 'runners', NAN, NAO 21, NCO to Sec. SWA, Ondangwa, July 5, 1939; NAO 20, Monthly Reports Ovamboland, May and Aug.-Sep. 1940. For elephants and border markers in western Ovamboland, see NCO to Sec. SWA, Ondangwa, 26 Aug. 1940. In 1949 ten elephant carcasses with the tusks removed were found in the uninhabited border area between Ovamboland and Kaokoveld, NAO 60, Quarterly Report Ovamboland, Oct. – Dec. 1949.

River, suffered from increased elephant raids on crop fields and attacks on cattle during the rainy season, and the destruction of water holes during the dry season.[74]

Predators, including lions, followed the large herbivore herds on their migrations. But when heavy rains turned the Ombuga Flats or the areas further west into swamps, prides of lions that regularly attacked the herds of wildebeest and zebra were cut off from the migrating herds when they moved south, and the lions found livestock on the edge of the inhabited zone to be an easier prey. Significantly, the lions that preyed on livestock near the villages during the 1940s and 1950s usually consisted of entire prides and not weakened old individuals.[75]

The eastern game migration corridor linked the Kavango / Cubango River Valley and Oshimolo with the Sandveld and ran as far south as Etosha Pan. Game that migrated from Etosha along the western game corridor also moved into the Sandveld.[76] Elephants moved along the entire length of the eastern game corridor, from Etosha to the Kavango River and between the eastern Sandveld north and northeast into Oshimolo. The elephants remained south of the border during the rainy season, but moved north and northeast and northwest (to Okafima and perhaps from there to the wilderness north of Evale and to the Kunene River) as the dry season progressed and surface water in pans in eastern Ovamboland dried out.[77] In eastern Ovamboland, the cattle post and settlement frontiers moved outward from the floodplain eastwards, while the eastern game migration routes ran on a north-south axis between Etosha and Oshimolo and the Kavango River. Contact and conflict between people and their livestock and elephants over water, food and space in eastern Ovamboland greatly increased when the settlement frontier met the eastern game migration corridor between Omboloka and Okongo during the 1930s, 1940s and 1950s.[78] Most of the

---

[74] NAN, NAO 20–21, Monthly and Quarterly Reports Ovamboland, Aug.-Sep. 1939, March – May 1940, Jan. – July 1941 and Jan. – June 1946.

[75] NAN, NAO 19–20, Monthly Reports Ovamboland, April 1931, June – Aug. 1939, Jan.-Sep. 1940 and NAO 60–61, Quarterly Reports Ovamboland, April – June 1949 and 1953–1954.

[76] NAN, NAO 19–21, Monthly Reports Ovamboland, 1931, 1939–1940, 1943, and NAO 60–61, Quarterly Reports Ovamboland, 1949 and 1953–1954.

[77] NAN, NAO 20, Monthly Report Ovamboland, May 1940; NAO 67, ANC to NCO, Oshikango, 11 July 1947; NAO 16, E.V. Martins [Veterinary Surgeon Lower Kunene] to Hahn [NCO], Lubango, 28 April 1945.

[78] NAN, NAO 20, Monthly Report Ovamboland, May 1940; NAO 67, ANC to NCO, Oshikango, 11 July 1947; NAO 16, Martins to Hahn, Lubango, 28 April 1945; NAO 67, ANC to NCO, Oshikango,

elephant damage to crops, water holes and dams as well as colonial border markers occurred in what was notorious as 'elephant country', between border markers 29 and 35.[79]

Animal attacks could be expected at remote cattle posts and pioneer settlements in the wilderness. Wild dog attacks on cattle posts occurred in 1928 and lion raids on cattle posts were reported in 1928, 1931, 1935 and 1940. At night, herdsmen at the cattle posts and along the trails leading to them drove the cattle into enclosures constructed of dense barricades of thorn bush that were taller than a man, and they kept fires burning throughout the night and remained vigilant by day and by night.[80] In April 1931, at least three serious lion attacks occurred. A heavy rainy season provided abundant browse and grazing, scattering wildlife over large areas.[81]

Evidence suggests that the incidence of predator attack on livestock in and near the villages increased during the late 1930s and 1940s after the colonial administration had disarmed the inhabitants of Ovamboland. The handing over of all modern firearms deprived Ovamboland's inhabitants of their most effective defense against wild animals, leaving them with a few antiquated muzzle loaders at best. Hahn became aware of the impact of disarmament because of a rapid increase in predator attacks in 1940. In Ondonga district, lions and wild dogs killed 52 cattle, 11 donkeys and a horse during a single week in September 1940 alone. At one location, Ondonga livestock owners lost 13 cattle and 6 donkeys. In the same year, lions killed several head of cattle and 4 precious horses in eastern Oukwanyama. The total livestock losses due to predator attacks in Ovamboland for the year were at least 100 head of cattle, 80 goats, 30 donkeys and 6 horses,

---

11 July 1947; ANC to NCO, Oshikango, 11 and 17 May, 1952; NAO 60, Quarterly Report Ovamboland, April – June 1952; NAO 17, NCO to Sec. SWA, Ondangwa 27 April 1933, and NCO to Sec. SWA, Ondangwa, 20 April 1935.

[79] NAN, NAO 67, ANC to NCO, Oshikango, 11 July 1947; ANC to NCO, Oshikango, 11 and 17 May, 1952, and NAO 60, Quarterly Report Ovamboland, April – June 1952. Cf. KAB 1, Volkmann, 30 Oct. 1928, "Report on the Agricultural and Political Conditions at The Angola Boundary", and NAO 18, Monthly Report Ovamboland, February 1927. On border markers and elephants, NAO 17, NCO to Sec. SWA, Ondangwa 27 April 1933 and NCO to Sec. SWA, Ondangwa, 20 April 1935. On competition for water, see NAO 17, O/C NAO Oshikango to NCO, Oshikango, 30 July 1934.

[80] See NAN, NAO 19–20, Monthly Reports Ovamboland, Sep.-Oct. 1935, Aug. – Nov. 1940, and NAO 105, Diaries kept by the NCO, Diary 1928 (stamped 1 Jan. 1928), entries 12 and 22 Nov. 1928. See also Timotheus Nakale, interview by author, Ekoka laKula, 21 Feb. 1993.

[81] NAN, NAO 19, Monthly Report Ovamboland, April 1931.

although the Native Commissioner acknowledged that it was impossible to give accurate numbers because the number of losses was underreported. The next year livestock losses to predators included 55 head of cattle, 60 goats, 20 donkeys and 4 horses. The losses were lower in 1942 but "[s]everal instances were reported where lions came right into the tribal area and in two cases natives were badly mauled". In 1943, reported livestock losses due to predator attacks were 12 head of cattle, 20 donkeys and 4 horses (mainly the prey of lions) and an unspecified number of small stock killed by smaller carnivores. The decline in the livestock attacks seemingly coincided with a collapse of the numbers of hyenas.

Many of the attacks took place in or very close to villages. In 1939, a hyena attacked two people in Ongandjera, killing one, and in the Oukwanyama district "what is presumed to be a rabid leopard attacked and badly mauled ten natives one of which has since died". Predators returned to haunt Ovamboland's villages during the late 1940s and early 1950s. Early in 1946, lions killed a village headman and severely wounded two or three other people in different villages on the edge of the districts' inhabited zones. Late in 1948, a leopard mauled a man right near the colonial administration's headquarters at Ondangwa in the densely settled heartland of the Ondonga district, and lions attacked cattle in various areas in April – June 1949. Early in 1953, "[a] wild dog which ran amok in the Ongandjera area attacked and bit twelve people, two of whom died of their wounds. The wild dog was killed".[82] The villagers defended themselves and their livestock by using poison, traps, bows and arrows and muzzle loaders, but these weapons were often ineffective when entire prides of lions were involved, and the colonial administration rarely made modern rifles available.[83]

Hahn refused to come to the aid of villages threatened by elephants and his successor Eedes only allowed a minimum number of rogue elephants to be shot. Hahn knew that elephants could cause severe damage. In May 1939 Hahn observed elephants very close to the Uukwaluthi tribal area "but up to

---

[82] NAN, NAO 36, District Surgeon Ovamboland, Monthly Report June 1939; NAO 19–21, Monthly and Quarterly Reports Ovamboland, April 1931, June-Sep. 1939, Jan. – May and Aug. – Nov. 1940, Jan. – June 1946. See also NAO 60–61, Quarterly Reports Ovamboland, Oct. – Dec. 1948, April – June 1949, Jan.-Sep. 1953 and Jan.-Sep. 1954; A450, 7, Annual Reports Ovamboland 1940–1943.

[83] NAN, NAO 20, Monthly Reports Ovamboland, Feb. 1940, March – April 1940 (poison); NAO 61, Quarterly Reports Ovamboland, Oct. – Dec. 1953 and 1954.

the time of writing they do not appear to have done any damage to crops or native food [storage] baskets". In August of the same year elephants moved into the tribal areas of Uukwaluthi and Onkolonkathi, but as the crops had already been harvested, according to the report "very little damage was done". In March – April 1941, elephants damaged a water hole in Ongandjera and destroyed crops in Uukwaluthi and Ondonga as water and food shortages increased. Hahn dryly remarked that without the aid of firearms people could not deal with the problem and that he lacked the staff and other resources to act. In 1942, elephants in the northwestern part of Ovamboland at various occasions entered homesteads, chased the inhabitants and ripped open the grain storage baskets to eat the millet reserves. Early in 1946, in Onkolonkathi district, a bull elephant "which was molested by the dogs of native cattle herds became infuriated and charged cattle which were grazing nearby. The bull struck a cow with its trunk and broke her back. An ox was picked up bodily and bashed into a mopani tree. Both the cow and the ox were killed on the spot". Towards the end of the rainy season in the same year, at least three elephants fell into water holes in Eunda and Uukwambi. In 1947, elephants wiped out the crops of sixteen households in six Eastern Oukwanyama villages; one elephant was shot. In Ondonga another two elephants "which were destroying water holes (...) were shot".[84]

The reluctance of colonial officials to act against the elephant incursions caused disbelief, fear and mounting anger. In 1949, the Ondonga King Kambonde reported that "I saw just 400 yards from my kraal a [*sic*] elephant getting in an old woman's kraal and pulling down one grain basket". Four elephants that entered the villages of Ondonga district damaged crops and water holes and were shot on orders of King Kambonde. Eedes with some reluctance allowed the headmen of Uukwambi and Ongandjera to shoot a few young elephants who had fallen into water holes. There were also reports of elephant poaching: at Ombombo deep into the wilderness south of Uukwaluthi, ten elephant carcasses with their tusks removed were found. In 1952, elephants caused problems in several districts and three were shot in eastern Oukwanyama. The Uukwaluthi king sent off a furious letter to Native Commissioner Eedes:

---

[84] NAN, NAO 20, Monthly Reports Ovamboland, May, Aug.-Sep. 1939 and March – April 1941; NAO 21, Quarterly Report Ovamboland, Jan. – June 1946; NAO 60, Quarterly Report Ovamboland, April-Sep. 1947; A450, 7, Annual Report Ovamboland 1942.

> You say that we the Ukualuthi people merely wish to kill the elephants (...) the elephants came to the Ukualuthi inhabited area at the edge of the Ukualuthi tribal area. There they met with cattle which were going to the bush. They killed four head of cattle out there. Our strength lies in the cattle (...). We will kill the elephants because they are killing our cattle. Now we have reported this matter to you on two occasions and you said the elephants do no harm in here.

In September 1953, elephants killed a blind woman, a child and an old man in Ongandjera district. In addition, elephants killed cattle, destroyed fruit trees and water holes, emptied out other water holes and water reservoirs and caused people to flee their homes. Only then did the Native Commissioner authorize shooting four elephant calves. At the same time, a large number of elephants roamed freely through Ongandjera. A herd of sixty elephants passed the Finnish Mission station. The official who wrote the quarterly report for the period merely commented that the elephant herd that terrorized the Ongandjera district "did not do any damage except to flog four goats to death".[85] At a 1958 annual tribal meeting in Ondangwa, Johannes Kuandambi asked if the Native Commissioner would now finally give people the permission to eradicate such "vermin" as lions and elephants.[86]

Government officials and local leaders continued to have very different interpretations of what constituted conservation. In 1971, the Director of Agriculture explained to the President of the South African Hunters and Game Conservation Association in Pretoria that the number of elephants in Ovamboland was at most about one thousand, and probably only a few hundred. For Ovambo farmers along the game corridors, however, elephants remained a plague especially during drought years. In the far east of Ovamboland, elephants continued to migrate across the border. In 1963, only two years after completion of a new border fence, elephants had destroyed it between markers 37 and 42

[85] NAN, NAO 20, Monthly Reports Ovamboland, Jan. – July 1941; NAO 60–61, Quarterly Reports Ovamboland, April-Sep. 1947, Jan-Dec. 1949, April – June 1952 and July-Sep. 1953; NAO 67, NCO to Sec. SWA, Ondangwa, 15 July 1947, and ANC to NCO, Oshikango, 11 July 1947; Chief Kambonde to NCO, Okaroko, 27 Dec. 1949; ANC to NCO, Oshikango, 11 and 17 May 1952; Chief Shetuatha Mbashu to NCO, Uukwaluthi, 12 Oct. 1952, and the tribal Secretary for Chief Shetuatha Mbashu to NCO, Uukualuuthi, 22 Sep. 1953.

[86] NAN, BAC 44, Minutes Annual Meeting Ondonga, 1 Dec. 1958.

at 44 different places.[87] In 1966, tensions about the conservation of predators and elephants in Ovamboland had escalated so much that the Commissioner General for South West Africa (South Africa's colonial governor for Namibia) wrote to the Minister of Bantu Administration (formerly Native Affairs) in Pretoria to request blanket authorization to shoot predators and elephants on sight when they entered inhabited areas, before they caused any damage. He emphasized that it was impossible to run Ovamboland and the other Northern Homelands as Native Reserves and game reserves at the same time, and recommended keeping the Department of Nature Conservation out of Ovamboland and the other homelands because it was greatly despised.[88] The Commissioner General's recommendations went unheeded and elephants and lions continued their attacks.

During the early 1970s, elephants caused damage to farms and water sources in eastern Oukwanyama, as well as in Uukwambi, Ongandjera and Uukwaluthi in the west, compelling the Director of Agriculture to write to the Secretary for South West Africa for permission to shoot two elephants in each of four different locations.[89] In May 1971, a rhino damaged a field in Oundhiya near Ukuangula.[90] In December 1977, elephants caused serious problems in Ongandjera and in the Ongandjera / Uukwambi border area. One elephant in Ongandjera killed five calves and a horse after two elephant calves had drowned in a well. Five elephants trapped in wells were killed. A rather insensitive Nature Conservation official reported that he had advised the villagers who had lost cattle "not to be angry because elephants also need to drink water".[91] In 1979, elephants were active in Uukwaluthi, Ongandjera and Ondonga "and broken fences and windpumps are a

---

87 NAN, AGR 95, Veterinary Inspector Omafo to Department of Agriculture Windhoek, Omafo, 20 June 1963.

88 NAN, BON 1, Commissioner-General SWA to Minister Bantu Administration and Development, [Windhoek], 21 Jan. 1966 (marked: "highly confidential").

89 NAN, OVA 54, Director Agriculture to President South African Hunters and Game Conservation Association (Pretoria), Ondangwa, 9 March 1971, and telegram Director Agriculture to Sec. SWA Windhoek, Ondangwa, n.d. [1970 or 1971?]. See also OVA 59, Monthly Reports [Agriculture], Leonard Haihambo, Oukwanyama, Haihambo to Oostehysen (Agriculture Ondangwa), Ohangwena, 18 Oct. 1972.

90 NAN, OVA 40, Student-Agricultural Official (J.P. Booysen), Travel Report No. 7/71, 21–25 June 1971.

91 NAN, OVA 54, Monthly Report Nature Conservation, Dec. 1977.

common sight".[92] In 1981, elephants caused great damage in the newly developed Mangetti area in the southeast of Ovamboland when they destroyed dams, pumps and water sources.[93]

Although Africa shared the Old World's disease environment, the continent proved vulnerable to invader species around the turn of the twentieth century, highlighting that Crosby's concept of ecological imperialism is as applicable in the case of Africa as it is to the Americas. But evidence in Africa also demonstrates the limitations of depicting biological invasion as a unilinear, mechanical and progressive process of environmental change from Nature to Culture (indigenous-natural to invasive-Western cultural). The impact of biological invaders and opportunistic indigenous species in north-central Namibia suggests the need for differentiating the process and the outcome of biological imperialism. Not all invaders had a significant impact, and the timing of their impact differed. The destructive impact of lungsickness and rinderpest in Africa mirrors that of smallpox in the Americas to the extent that the epizootics caused immediate and dramatic domestic and wildlife losses, weakening preconquest societies and the environments they depended upon, and in the process paving the way for colonial conquest. Lungsickness is a critical factor in explaining the collapse of South Africa's Xhosa livelihoods and society and its subsequent conquest. Rinderpest had an enormously destructive impact across southern Africa in 1896 and 1897, bringing many communities to the brink of collapse. Unwilling or unable to eradicate the diseases, colonial administrations cordoned off the infected domestic and wild animal herds. Thus, invasive germs that accompanied the Europeans in fact turned against them by impeding progress on the grand colonial development plans that were based on scientific livestock management and the commoditization of indigenous cattle.[94]

Disease—caused by invasive and local microbes—may have been a much more critical factor in the conquest and colonization of Africa than has been generally accepted. It did not result in a demographic collapse of a magnitude comparable

---

92 NAN, OVA 43, Secretary for Agriculture, Oshakati, 13 Sep. 1979.

93 NAN, OVA 33, Memo Department of Agriculture (Ovambo Administration) to Director Department of Agriculture and Forestry, Ondangwa, 8 May 1981.

94 For a good example of an invasive plant species turning on its fellow invaders, see van Sittert, "'The Seed Blows About in Every Breeze'".

to that in the Americas. The most destructive impact of disease on human populations was indirect, as with lungsickness, rinderpest and foot and mouth. Yet their impact was catastrophic because of the extent to which people's lives and livelihoods depended on the (domestic) animals affected by the diseases. Lungsickness and rinderpest decimated cattle resources. But, again, the impact of diseases in Ovamboland defies linear and mechanical models of environmental change. Rinderpest and lungsickness caused severe animal losses, but foot and mouth did not do so, even as the colonial administration used draconian measures to contain it.

The history of donkeys in north-central Namibia also defies unilinear progressive models of biological invasion. Unlike sheep in northern Mexico, donkeys initially did not thrive in the region at all and their introduction does not adhere to the invasion-explosive growth-implosion model. The case of biological imperialism in Ovamboland thus raises the same questions about the unilinear Nature-to-Culture depiction of environmental change and the homogeneity of the process and its outcome that emerged in the previous chapter in regard to imperial political ecologies.

5

Guns, hoes and steel: Techno-environmental determinism

Technology can serve as a catalyst for environmental change but again the process is usually cast in very linear and mechanical terms.[1] Once the process is set in motion, further technological innovation follows, ultimately resulting in the substitution of the natural by the artificial and, ultimately, of Nature by Culture. Diamond has argued that because Europe benefited from certain geo-environmental advantages, the continent led the way to acquiring guns and steel technology. According to Diamond, these advantages explain why the West colonized the non-West, rather than the other way around. Diamond thus offers an environmental explanation (the geographical realities of the European environment facilitated borrowing) for historical phenomena (European imperialism and colonialism) that had transformative environmental impacts locally and globally. Western technology was instrumental not only in paving the way for military and political conquest. It also facilitated environmental conquest within a Nature-Culture dichotomous framework: technology first allowed the West to convert its own environment from Nature into Culture, and subsequently provided the tools with

[1] For a strong critique of linear technological determinism, see Thompson, *The Soundscape of Modernity*.

which to conquer non-Western environments. In this model, the main obstacle to Western dominance is not really non-Western *Culture*, but rather non-Western *Nature*, because most non-Western societies were held to be both hostage to and part and parcel of their physical environments. The West thus could not conquer Africa militarily and politically until it had developed the necessary technology (i.e., Western medical science and quinine) to overcome the continent's natural disease environment. Further advances during the post-World War II era made it possible to address the last remaining Natural obstacles to Western development in Africa by using techno-science to eradicate animal and human diseases and to introduce Western technology (erosion control and irrigation, plows, improved cultivars) to harness Africa's natural resources and effectively exploit them.[2]

Whereas Diamond's guns and steel argument appears at first glance persuasive for the fifteenth- and sixteenth-century Americas, since Amerindians had neither steel nor guns with which to resist European conquest, the situation in Africa was radically different. Africans produced and used steel weapons and tools long before Europeans, and from the sixteenth century onwards, Africans rapidly acquired firearms that enabled them to resist European imperialism until the end of the nineteenth century. While Europeans conquered and colonized much of the Americas between the late fifteenth and the late nineteenth centuries, at the end of the nineteenth century, Europeans still controlled precious little territory in Africa.

Guns

In *Guns, Germs and Steel*, Diamond identified guns as weapons that gave Europeans a critical military advantage. But how critical were guns, really? Cortés and Pizarro had few guns and the impact of guns may have been more in terms of morale because guns were not very effective as weapons. Moreover, Amerindians and Africans acquired guns during the wars of conquest in the Americas and in Africa.

---

[2] See chapters 4 and 6.

In many cases, guns not only facilitated Western conquest, but they also have delayed, prevented or reversed it.[3] Modern small arms, notably the Kalashnikov AK-47, are credited with liberating the colonized world during the anticolonial and anti-imperial struggle. Mozambique, which emerged as an independent nation after a protracted liberation war in the 1960s and 1970s, prominently bears the image of the Kalashnikov on its national flag.

Although Europeans initiated and controlled the overseas trade in slaves as well as their use in the Americas, they did not control the raiding and the supply of slaves in Africa itself. One explanation for this is that through the gold, slave and ivory trade during the eighteenth century some West and West-Central African polities acquired enormous quantities, often of the latest technology of guns and ammunition. The same was true in the nineteenth century not only in West and West-Central Africa but also in southern and eastern Africa. In terms of small arms, Africans often were as well armed as the Europeans. Around the turn of the eighteenth century, the Asante purchased large quantities of guns and gunpowder from the Dutch West India Company. In 1706 alone, the Director-General of the West India Company on the Gold Coast requested 6,000 guns and 100,000 pounds of gunpowder from the Netherlands to meet demand there. The guns were the latest flintlocks rather than the outdated matchlock guns that were still in use in Europe at the time.[4]

Africans also acquired a not inconsiderable amount of the latest gun technology in the late nineteenth century: first percussion muzzle loaders and rifles, then single-shot breechloaders and magazine-fed repeaters, and finally bolt-action rifles. Samory's forces in West Africa had modern rifles and blacksmiths who could build rifles, although repair and manufacture of firearms was typically a handcraft activity rather than a mass-production one as in Europe. Still Samory's several hundred blacksmiths produced twelve breechloaders per week as well as 200–300 cartridges. The Boers of South Africa also depended on imported firearms, but often were equipped with better small arms than their British

---

[3] Headrick highlights how guns facilitated conquest, see *The Tools of Empire*, pp. 83–126.

[4] Kreike, "Early Asante and the Struggle for Economic and Political Control on the Gold Coast". Kea argues that the gun replaced the bow as the main weapon of Gold Coast armies around 1700, in *Settlements, Trade, and Polities in the Seventeenth-Century Gold Coast*, pp. 156–164.

opponents in the South African War of 1899–1902. Breechloaders and repeaters flooded the international and African markets in the aftermath of the American Civil War and the Franco-Prussian War.[5]

Even though firearms had been unknown before the 1860s, by the late nineteenth century the northern Ovambo floodplain polities of the Ombadjas and Oukwanyama were well supplied with guns. King Moshipandeka Hepunda (1862–1882) of Oukwanyama had 1,500 firearms that were mostly muzzle loaders, but also included 50 breechloaders. His successor, King Namadi Mwelihanyeka (1882–1884), had a guard 3,000 men strong, armed with Martini-Henry and Westley-Richards breechloading rifles. By 1910, Oukwanyama alone had 15,000 firearms, of which 7,000 were breechloaders, including American magazine-fed Winchester repeating rifles and German Mauser bolt-action rifles. The Ombadja kingdoms west of Oukwanyama were also well supplied with modern firearms.[6]

The southern Ovambo floodplain polities were less wealthy in ivory and cattle, and so less able to purchase the most advanced firearms, but when the South African colonial administration disarmed their inhabitants in the mid-1930s, the number of rifles (i.e., breechloaders) that were confiscated totaled almost 3,500, including over 2,000 in Ondonga, close to 1,000 in Uukwambi, 140 in Eunda and Onkolonkathi and 182 in Ongandjera. In 1933, after a series of firefights between different factions, the Onkolonkathi headman handed in 127 modern rifles, most of which were magazine-fed, eliciting Native Commissioner Hahn's comment: "This small tribe was much better armed than was thought". Over 3,000 of the confiscated rifles were kept in storage at the administrative offices at Ondangwa until they finally were destroyed in 1947.[7] Thus, during the 1900s and 1910s era of colonial conquest, the Ovambo polities jointly could have assembled at least 10,000 men armed with breechloading rifles. The Portuguese army that invaded Oukwanyama in 1915—the largest army that had been fielded until then

---

[5] The Boers imported modern firearms from Europe on the eve of the South African War, see Thompson, *A History of South Africa*, p. 141. On Samory and modern firearms, see Person, *Une révolution Dyula*, 3:1762; Kanya-Forstner, "The French Marines and the Conquest of Western Sudan", p. 136. For the postconflict surplus of small arms and Samory's manufacturing of firearms, see Headrick, *The Tools of Empire*, pp. 109–110 and 119–120.

[6] Kreike, *Re-creating Eden*, pp. 27–23 and 41–42; Capello and Ivens, *De Angola a Contra-Costa*, p. 228.

[7] NAN, NAO 20, Annual Reports Ovamboland, 1937–1938, and Monthly Reports Ovamboland, June 1933 and Dec. 1939; NAO 60, Quarterly Report Ovamboland, Jan. – March 1947.

by the Portuguese in Africa, consisted of 10,000 men who were armed with breechloaders, machine guns, and cannon. Challenged by Oukwanyama, armed with approximately 7,000 breechloaders and a similar number of flintlock and percussion cap muzzle loaders, the Portuguese army came close to defeat.[8] For comparison, in the 1879 Anglo-Zulu war a British army of 18,000 and 9,000 African levies faced 29,000 warriors who may have been armed with up to 20,000 firearms, mostly old muzzle loaders, and only 1,000 modern breechloaders. The Ethiopian army that in 1896 crushed the Italian army at Adowa had several thousand breechloaders while Samory's army in 1898 had 4,000 repeaters.[9]

The inhabitants of the Ombadjas and Oukwanyama captured large amounts of small arms and ammunition, as well as several machine guns and cannon, from their Portuguese opponents following a string of Portuguese defeats. An Oukwanyama blacksmith even repaired one of three captured cannons, a French 75; the Portuguese had removed the breechblock locks from the cannons to prevent them from being used against them, but the blacksmith forged a replacement lock that rendered the gun operable. When the Portuguese recaptured the cannon, "the Commanding Officer was so impressed with the native made part that he sent it to the leading museum in Lisbon". An Anglican missionary identified the blacksmith as Twemuna but Native Commissioner Hahn claimed the feat had been accomplished by Hangush, a headman who died in 1940.

A large majority of the firearms had been acquired from European traders and smugglers at high prices. In 1870, a flintlock muzzle loader in southern Angola cost 10–15 head of cattle. By 1879, availability of and demand for more advanced firearms dropped the price of flintlocks to one head of cattle, and the Angola-based trader Chapman reported that the inhabitants of the Ovambo floodplain refused to buy flintlocks, preferring more modern guns. At the time, the more reliable percussion cap muzzle loaders cost two head of cattle, while such advanced breechloaders as the Martini-Henry and the Snyder were traded for a staggering 48–54 head of cattle apiece. In the 1890s, the percussion cap muzzle loaders had been reduced in price by half, whereas advanced breechloaders cost 7–10 head of cattle apiece. In Central Sudan in 1895, a Martini-Henry cost

[8] Kreike, *Re-creating Eden*, pp. 52–53.
[9] Taylor, *Shaka's Children*, pp. 216–217; Headrick, *The Tools of Empire*, pp. 119–120.

13–33 slaves and in 1898 Somalia, a rifle cost 5–6 she-camels.[10] The acquisition of guns, which in the context of rapid technological advances in firearms technology and the threat of colonial expansion and conquest reached the proportions of a veritable arms race, constituted a serious drain on Ovambo floodplain resources. Cattle were the main export commodity for purchasing firearms, ammunition and horses, and, particularly after the rinderpest decimated cattle populations in the region, conflict over cattle increased, leading to escalating violence, population displacement and further investment in guns for defense against cattle raiders.[11] In 1890s prices, the 10,000 breechloading rifles the Ovambo acquired in the early 1900s cost the equivalent of 70,000 to 100,000 head of cattle, whereas colonial estimates for the total number of cattle for Ovamboland were 60,000 in 1925 and 123,000 in 1942.[12]

Thus firearms had contradictory effects: not only did they facilitate the European conquest, but they also helped prevent (in the case of Ethiopia) or delay European conquest and colonization. The impact of guns as a technology was therefore neither linear nor homogenous. The resistance of Samory, the Ethiopians, the Boers and the northern Ovambo floodplain polities of Oukwanyama and Ombadja would not have been as effective or as sustained without modern firearms. But being met with effective resistance also redoubled Portuguese determination to conquer the area at all costs to maintain legitimacy as a victorious imperial and colonial power. The incursions of well-armed Oukwanyama and Ombadja raiding parties also increased violence and insecurity throughout southern Angola, further challenging the Portuguese.

The South Africans disarmed the refugees from Oukwanyama and the Ombadjas on their side of the border in 1917 after they defeated Mandume Ndemufayo, the last king of Oukwanyama.[13] The colonial administration allowed selected kings and headmen to keep a small number of modern firearms—100 in the case of Ondonga—but required all firearms to be registered. It appears that some leniency was extended to muzzle loaders. A couple of hundred barrels without

---

[10] Headrick, *The Tools of Empire*, p. 110.

[11] Kreike, *Re-creating Eden*, pp. 26–32, 41–47. On the refusal to purchase flintlocks, see NAN, A233, J. Chapman, 1903–1916, pp. 18–21. On Hangush, see A450, 7, Annual Report Ovamboland 1940.

[12] Kreike, "Architects of Nature", p. 112, table 6.1.

[13] See chapter 3.

stocks (probably muzzle loaders) were handed in in 1934, but at least some muzzle loaders appear to have continued in use. The administration distributed small numbers of cartridges for the modern guns to the kings and headmen who had kept them. Although gun smuggling had been rampant before colonial occupation during World War I, it seems to have continued at a much reduced scale during the 1920s, and it had become virtually nonexistent by the 1930s. As late as 1928, three "runners" for a Portuguese dealer known as Mengeri from the Humbe area in Angola were arrested: they carried 40 percussion caps, three tins of gunpowder and a Mauser pistol with 69 rounds. In addition, colonial records occasionally note that returning migrant laborers carried a few rounds of ammunition for the registered firearm of a particular headman.[14] In 1930, the trader Hermann Burchard, a resident of the Angolan Lower Kunene and a former inhabitant of the German colony of South West Africa, was banned from re-entry into Ovamboland because he was suspected of having supplied rifles and ammunition to King Iipumbu of Uukwambi in 1925.[15] During World War II, the colonial administration issued 36 .303 rifles (probably war-issue Lee-Enfields) to a selection of Ovamboland's headmen.[16]

Combined with the prohibitions on Africans hunting big game—one of the reasons given for the violent removal of King Iipumbu in 1932 was that he had engaged in 'poaching'—the disarmament of the Ovambo in part explains the resurgence of big game populations in the 1930s. The introduction of breechloading cartridge-fed rifles (especially magazine-fed ones) had greatly enhanced the efficiency of big game hunting in the Ovambo floodplain during the close of the nineteenth century, leading, for example, to the near disappearance of elephants from the region. The lack of firearms and cartridges also reduced the impact of hunting big game. The resulting recovery of wildlife populations caused increased conflict between predators and elephants on the one side and humans on the other. Colonial reports demonstrate that the local population often was armed with spears, bows and arrows and muzzle loaders, making hunting less effective and

---

[14] NAN, NAO 26, Acting UGR Neutral Zone, Namakunde, March 18, 192[7] to Portuguese Govt. Representative Neutral Zone, Namakunde (attached to letter Acting UGR Neutral Zone, Namakunde, Feb. 25, 1928 to O/C NAO, Ondonga). In general, see NAO 18, Annual Reports Ovamboland 1928, 1930, 1932. On precolonial gun smuggling, see Kreike, *Re-creating Eden*, pp. 42–43.

[15] NAN, NAO 26, NCO to Secretary SWA, Ondangwa, 24 July 1929, and 5 Dec. 1930.

[16] NAN, NAO 60, Quarterly Report Ovamboland, Jan. – March 1947.

much more dangerous. The 1931 annual report, for example refers to a "miniature war" with a pride of lions that had killed two donkeys on the edge of the Ondonga inhabited area: some of the hunters were armed with muzzle loaders "and after discharging their weapons [they] were forced to beat a retreat in order to re-load". The hunters killed two of the four tracked lions, but not until one of their own had been badly mauled. In Uukwaluthi lions attacked a herd of livestock but, added Hahn, "I have not heard yet whether any natives were shot," demonstrating that he had not only little respect for the marksmanship of his subjects, but also no sympathy for their safety or their attempts to protect their precious livestock.[17]

Moreover, it was difficult to obtain supplies of gunpowder, shot and percussion caps or flints, or spare parts for these guns, which were at least forty to sixty years old. The disarmament of the Ondonga district was only completed in 1939, but Native Commissioner Hahn noted the unexpected consequence almost immediately. In his 1940 annual report he wrote: "The Ondonga natives have suffered considerably since they were disarmed last year. This tribe is bordered by the [Etosha] Game Reserve on the East as well as on the South and as a consequence lions, hyenas and wild dogs affect them much more closely than any of the other tribes". In the next annual report he also stressed that because the inhabitants of Ovamboland had few rifles left they could not effectively keep the lions at bay. Yet, Hahn hesitated to respond to a request by the headmen to issue them rifles for defending themselves against the predators as the attacks mounted.[18]

Forced disarmament in Ovamboland and the rebounding wildlife population may also have been important reasons that the local population continued to construct elaborate palisades around their homesteads and fences around their fields, even though levels of violence declined in the 1930s. The defenses provided effective protection for humans, their livestock and their crops and fruit trees against big and small game depredations. Disarmament may have contributed to big game conservation in north-central Namibia, but the policy simultaneously may have maintained or even increased the use of woody vegetation as an alternative protection against wildlife.

---

[17] NAN, NAO 19, Monthly Report Ovamboland, April 1931.

[18] NAN, A450, 7, Annual Reports Ovamboland 1940 and 1941.

Steel tools

A second technological advantage that Diamond identifies as being critical for the success of European imperialism was iron technology and steel. In sixteenth-century warfare in the Americas, Cortés' and Pizarro's opponents did not possess metal arms: at best they used obsidian for cutting edges while the Spaniards had iron swords, pikes, bolts and arrowheads, as well as protective iron helmets, shields and body armor. But Africans used the same metals as Europeans. The oldest sites of iron working in sub-Saharan Africa date from the ninth to eighth centuries BC. Iron technology not only had an enormous impact on warfare (swords, knives, axes, spears, arrowheads) but also on agriculture because metal bush-knives / machetes, axes and hoes greatly facilitated agriculture, by making it easier to clear bush and forest, and to till, weed and harvest arable fields. In addition, iron smelting and working consumed large amounts of (hard)woods as a fuel.[19]

African blacksmiths produced steel weapons, tools and other artifacts long before the Europeans did. This explains why—despite the import of European raw iron and iron tools and arms since the seventeenth century, African indigenous iron smelting and working did not disappear until the first half of the twentieth century. Although sub-Saharan Africans may or may not have invented iron working independently, there is no doubt that it was a large-scale activity throughout Africa.[20] Late nineteenth-century travelers to thc Ovambo floodplain marveled at the iron industry in the Kingdom of Oukwanyama. The ore was mined in the Mupa area, north of the floodplain in modern Angola, and it was collected during annual expeditions.

---

[19] Iliffe, *Africa*, pp. 33–34. See also de Maret and Thiry, "How Old Is the Iron Technology in Africa?", Coucher and Herbert, "The Blooms of Banjeli", and Schmidt and Avery, "Complex Iron Smelting and Pre-Historic Culture in Tanzania". Denevan argues that in the Americas forest clearance and shifting cultivation only really became possible with the introduction of iron technology, see Denevan, *Cultivated Landscapes of Native Amazonia and the Andes*, pp. 27–49 and 116–123.

[20] Herbert, *Iron, Gender, and Power*, pp. 4–11. Coucher and Herbert note that many Africans believed that locally produced metal hoes and axes were stronger and more durable than imported tools. See Coucher and Herbert, "The Blooms of Banjeli", p. 50. Craddock emphasizes that the preference for tools produced by the unique iron-working technologies of precolonial Africa was not surprising because it resulted in high-quality iron with a high carbon content, i.e., steel. See Craddock, *Early Metal Mining and Production*, pp. 241–253, 264–265.

An authoritative study on African iron working explains that by the early twentieth century, African blacksmiths chiefly used (European) scrap metal as a resource rather than smelting iron themselves, that blacksmithing was moribund by the 1920s, and that after World War II, blacksmithing was limited to a few remote areas of Africa.[21] Still, while blacksmithing certainly declined rapidly in north-central Namibia after World War II, Oukwanyama blacksmiths on both sides of the Angolan-Namibian colonial boundary acquired raw iron from the Mupa region until the 1940s, and they remained an important source for hoes and other tools despite the availability of industrially produced hoes from Portugal and South Africa. For example, when Kaulikalelwa Oshitina Muhonghwo married in the early 1940s, her husband, the son of the village headman of Ohamwaala (in the Namibian district of Oukwanyama), paid her family the set bride wealth of one ox and four hoes. The hoes were especially valuable because they had been crafted from iron ore obtained from the Mupa area manufactured by Oukwanyama blacksmiths. She explained that when European-manufactured hoes were substituted for the local superior-quality variety, the equivalent for one 'Ovambo' hoe was two and a half imported hoes.[22] The missionary-ethnographer Carlos Estermann noted in 1935 that the customary bride price amongst the Ovambo consisted of a head of cattle and four 'indigenous' hoes.[23] In 1938, the official bride price in Ovamboland's Oukwanyama district consisted of one head of cattle payable to the father of the bride and six hoes payable to her mother.[24]

In 1935, a group of blacksmiths demonstrated iron smelting and the production of metal artifacts at an agricultural and ethnographic show in Windhoek. Native Commissioner Hahn had sponsored and organized the Ovamboland section of the show, and he had ordered a load of authentic iron ore from Mupa. He considered iron smelting to be an antiquated craft with no economic significance, and consequently offered a mere symbolic price of one shilling for the entire load that had been carried to Ovamboland by a caravan of thirty men. The convoy's leader, the Angolan blacksmith Shiweda, however, demanded ten shillings. Consulted by Hahn, the old Oukwanyama blacksmith Mueshipakange Hamuandi, who per-

21 Herbert, *Iron, Gender, and Power*, pp. 11–12.

22 Kaulikalelwa Oshitina Muhonghwo, interview by author, Ondaanya, 29 Jan. 1993.

23 Estermann, "Notas Ethnográficas", 55–56.

24 NAN, NAO 9, O/C Native Affairs to NCO, Oshikango, 22 Feb. and 27 April 1938.

formed at the Windhoek show, noted that the artifacts produced with such expensive ore would not fetch even half the requested price. Hahn refused to pay Shiweda's price and may have procured iron ore in the Namibian Police Zone, south of Ovamboland, instead. When he informed his superior about the show, Hahn claimed that the ore had originated in the Mupa area: "Iron ore, specially brought from the iron mines at Oshimanya in Angola (. . .) was smelted and forged at their primitive blast furnaces, and the metal fashioned into hoes, knives, assegais, axes, etc. (. . .) In many instances the anvil consisted of a big hard stone". It is unclear whether Hahn coldly lied to his superior, or whether he ultimately paid Shiweda's price for the Angolan ore.[25]

Ovamboland's pavilion at the Windhoek show featured six regular blacksmiths from (Namibian) Oukwanyama: Mueshipakange Hamuandi, Hikela Hidinua, Mateus Nakalondo Hamnyela, Shiteni Kamati, Himuthe Valombola Kakende and Deumane Kaushola. In addition, seven "jeweler-blacksmiths" participated: Haupindi Katinda, Hikushi Amnyela (nicknamed 'Bones'), Gepata Hamatundu, David Nafine or Hamaulu, Lieinge Kambode, Mu(a)etako Knghone and Hamnyela Haimbili. The regular blacksmiths made weapons and tools, while the jewelers produced anklets and iron- and copper-beaded jewelry.[26] Sometimes, jeweler blacksmiths were San men who specialized in manufacturing iron beads.[27] In 1936, Hahn organized a display of arts and crafts for the benefit of the visiting Administrator of South West Africa in Oukwanyama that "included blacksmiths, making hoes, axes, knives, assegais, etc., jewelers making beads, copper and iron bracelets, and brassware".[28] Hahn may have engaged the same blacksmiths that he had used for the Windhoek show in the previous year. An Anglican missionary stationed in the Namibian district of Oukwanyama during the 1930s and early 1940s witnessed five blacksmiths at work at the Native Commissioner's office,

---

[25] NAN, NAO 27, O/C Oshikango to NCO, Ondangwa, 15 March 1935; NAO 27, NCO to Sec. SWA, Ondangwa, 15 July 1935.

[26] NAN, NAO 27, NCO to Whom it May Concern (Pass) & Attached List for Windhoek Exhibition 1935, Ondangwa. 4 April 1935, and NCO to Sec. SWA, Ondangwa, 15 July 1935. The blacksmith Mateus Nakalondo Hamnyela may be the same as the village headman (of Oneleiua?) Matheus Hamnyela who in 1947 was a close associate of Senior Headman Vilho Weyulu and accused of employing forced labor, see NAO 98, anonymous to Ohamba Nakale [Eedes], received at Ondangwa 17 Jan. 1949, and ANC to NCO, Oshikango, 20 Jan. 1949. If this is the same individual, blacksmithing was unlikely to be have been his main occupation any longer.

[27] Kaulikalelwa Oshitina Muhonghwo, interview by author, Ondaanya, 29 Jan. 1993.

[28] NAN, NAO 19, Annual Report Ovamboland 1936.

transforming muzzle-loading gun barrels into hoes. The barrels had previously been used as pipes for distilling liquor. The missionary was quite impressed by the blacksmiths' skill, and he recounted the story of the famous Oukwanyama blacksmith who had repaired a captured Portuguese cannon.[29]

After World War II, however, blacksmiths in Ovamboland rapidly became rare. In 1948, the Native Commissioner reported that few blacksmiths remained and that their number was dwindling, especially in Oukwanyama, which had been the principal center for blacksmithing. The Commissioner explained: "they cannot compete with similar articles of European manufacture which now[a]days are more easily and cheaper acquired at the local shops".[30] During the early 1940s, Native Commissioner Hahn noted that the production of local knives decreased because of what he perceived to be a drop in demand rather than in supply: "the manufacture of the dangerous Ovambo sheathed knife is definitely on the decline. This is principally due to the fact that the territory has, as a whole, become much more peaceful".[31] Across the border in Angola, Carlos Estermann noted that in 1935, the Lower Kunene had many blacksmiths who still obtained their ore from Mupa, but that the extraction of iron from local ore was becoming increasingly rare.[32]

In Ovamboland, imported iron hoes and other European goods could only be obtained from a few shops, from the missionaries, from Portuguese traders across the border or from central or southern Namibia. From 1925 to 1939, the Ondonga Trading Company (OTC) enjoyed a monopoly over trading in Ovamboland as a private concession. The OTC maintained an outlet at Ondjondjo at Ondangwa from 1925 onwards, and soon thereafter opened a second store at Omafo in the Oukwanyama district close to the Angolan-Namibian border. During the early 1930s, the economic depression with the resulting decline in migrant labor wages in Ovamboland caused a slump in sales because people had "very little ready cash and are not inclined to spend (…). trade in Angola is as bad if not worse than in S.W.A.". By the end of 1932, business declined so sharply that the OTC closed its shop in the Oukwanyama district. By the middle of 1933, business

29 MacDonald Diary, p. 9.
30 NAN, NAO 71, NCO, Purchase and Sale of Native Curios and Handicraft, 20 Aug. 1948.
31 NAN, A450, 9, Annual Report Ovamboland 1941.
32 Estermann, "Notas Ethnográficas", pp. 51–52.

had deteriorated even further although the OTC shop at Omafo (Oukwanyama) had been re-opened by R.S. Cope, the effectively unemployed labor recruiter in Ovamboland, who was happy to earn at least some income through his small store, where the gross earnings amounted to 100 pounds sterling per month. His brother-in-law, Native Commissioner Hahn, noted, "there is always more ready cash in Ukuanyama [Oukwanyama] than in any of the other tribes. Besides this many Portuguese natives from Angola come over to trade at Omafo".[33] In 1939, the Ovamboland administration issued a license to the private trader Erich Beersmann to open a store at Endola in the Oukwanyama district because of the "excessive" prices that were being charged by the monopoly OTC.[34]

The two OTC shops and Beersman's shop offered a variety of metal goods. According to a 1940 report comparing the prices in Beersman's Endola store and the OTC ones, all outlets charged one shilling for hoes. Beersmann also carried a second type of hoe which was more expensive, ranging in price from 2s. to 2s. 6d. In the light of the above oral history account of the higher prices commanded by locally manufactured hoes the second type may have been an Ovambo hoe, although its larger size and its being listed as "Etema lombruru" (*etema lombulu* or Boer hoe?) suggest that it may have been a South African manufacture. Bush knives—imports since they were not produced by Ovambo blacksmiths—varied in price from 1 to 2s. at Beersman's Endola shop to 2s. at the OTC shops. An ax varied from 2s. 3d. to 4s. at Endola and 2s. to 4s. at the OTC outlets. The OTC also carried large axes and spades, and both the OTC and Beersman carried knives, nails and wire but these items were all cheaper at Endola. Both the OTC and Beersman also sold 'kaffir pots' (locally manufactured clay pots) but only the OTC had 'kitchen pots', which probably were imported metal pots, in its inventory.[35]

During World War II, prices at the shops increased while the availability of imported iron manufactures decreased. In 1941, the prices for regular hoes at the Endola store had increased to 1s. 6d., and the price of the second type of hoes

---

33 NAN, NAO 19, Monthly Report Ovamboland, Jan. and Sep.-Oct. 1932, June 1933.

34 NAN, NAO 25, Additional NC, Trading in Ovamboland: Ondonga Trading Company Limited, Windhoek, 23 Sep. 1938, Administrator SWA to NCO, Windhoek, 24 June 1939 and CNC to NCO, 6 Oct. 1938.

35 NAN, NAO 25, Erich Beersmann to NCO, Endola, 2 Nov. 1939, and Pricelist Ondonga Trading Company Ltd. (OTC), Jan. 1, 1940.

sold by Beersmann had risen to 2s. 6d. No prices for bush knives were given and it is likely they no longer were available in the shops. 'Kaffir' pots ranged in price from 7s. to 12s. With the cash infusion resulting from the remittances from military service wages in 1942, business at Ovamboland's shops boomed, but the stores had so much difficulty in acquiring new supplies of goods that a new 'native store' at Ombalantu was forced to close down almost as soon as it had opened because its owner could not procure inventory. Although the three remaining shops continued to do good business, many potential clients crossed the border to buy from the well-stocked Portuguese-owned shops in the Lower Kunene which were supplied especially with soft cotton goods from Brazil. In the final years of the war, metal goods were in short supply. The OTC, which in 1943 was taken over by the South West Africa Native Labour Association, the main migrant labor recruiting company, did not supply any hoes, axes, knives, wire, saws or (metal) kitchen pots. Only in 1947, two years after the end of World War II, did the stores once again carry a range of metal goods. But hoes cost 2s. (twice their prewar price) and bush knives and metal cooking pots were unavailable. The only cooking pots in the inventories were 'kaffir pots'.[36]

Diamond's 'steel' thesis, like his 'guns' thesis, explains why it was that before and during European-African contact, at the turn of the twentieth century, Europeans colonized Africans rather than the other way around. In contrast to the situation in the pre-Columbian Americas where iron working technology was nonexistent, Africans had used steel tools long before Europeans, and their indigenous handcrafted iron and steel hoes (and axes) had greatly facilitated agriculture and the clearing and utilization of woodlands for up to three millennia before the European colonial invasion. In north-central Namibia, indigenous blacksmiths and their handcrafted steel hoes held their own against industrially manufactured imports until well into the 1940s, defying linear narratives of the automatic substitution of 'inferior' traditional and local technology by 'superior' modern Western global technology.

---

[36] NAN, NAO 25, OTC Price Lists, 1 April 1941 and 26 Jan. 1944; A450, 7, Annual Report Ovamboland 1942 and 1943; NAO 21, Quarterly Report Ovamboland, Jan. – March 1943; NAO 63, Comparative List of Some Articles: Omafo and Endola Stores, appendix to NCO to CNC, Ondangwa, 15 Oct. 1947 and Price List SWANLA Ondonga and Omafo Stores, 16 June 1947, appendix to Store Manager SWANLA Ondonga to NCO, Ondonga, 16 June 1947.

Steel plows

The history of the plow in the Ovambo floodplain and southern Africa shows a similarly asymmetrical and contradictory history. The use of the animal-drawn plow is ancient in highland Ethiopia.[37] However, the plow did not readily disseminate to other parts of sub-Saharan Africa. During the nineteenth century, the use of the animal-drawn metal plow spread in the eastern part of southern Africa (e.g., Lesotho, Swaziland). In Ovamboland, the plow led to more labor-intensive preparation of the soil, but it also used more land to produce the same quantity of crops.

The new South African colonial rulers in Namibia first introduced the plow soon after they occupied the southern Ovambo floodplain in 1915. Yet, plows were not adopted in Ovamboland until after World War II, more than three decades later. Why did it take more than a generation for the animal-powered steel plow to be adopted in an area that had a strong history of crop cultivation, that was entirely free of tsetse and that had an abundance of cattle?

When the South African official based at Ondangwa first used a plow in 1916, three of his local employees requested his assistance in acquiring plows for themselves. In 1918, with the blessing of his superior in Windhoek, he ordered three plows from a South African import firm based in Port Elisabeth for seven pounds sterling each.[38] An agricultural expert for the Anglican mission, however, noted in 1924 that plowing would not really be effective in Ovamboland because tree stumps and roots routinely were left in the fields.[39] Even in 1938, it appears that little or no use was made of plows: the Council of Headmen of the district of Oukwanyama submitted a request for three "single furrow ploughs suitable for ploughing sandy loam soil".[40] The plows were first used during the agricultural season of 1938–1939. The new colonization area east of the Ovambo floodplain, where the overflow of refugees from Angola had been directed, received two of

[37] McCann, *People of the Plow*.

[38] NAN, RCO 7, Office of the Secretary for the Protectorate SWA to RCO, Windhoek, 1 Dec. 1917; Managing Director Mangold Brothers Ltd. to RCO, Port Elisabeth, 26 April 1918; and Administrator SWA to RCO, Windhoek, 23 May 1918.

[39] NAN, NAO 26, Report Ovamboland Cotton Prospects, appendix to Alec Crosby to Bishop of Damaraland, St. Mary's Mission, 11 Jan. 1924.

[40] NAN, NAO 11, NCO to Sec. SWA, 24 Aug. 1938.

them, which were financed through the Tribal Trust Fund of the Oukwanyama district. Senior Headman Eliah Weyulu had cleared a number of 'tribal' fields in the villages of Ohauwanga Munene, Omundaunghilo and Eenhana that were successfully plowed. The fields yielded a good harvest, but the results with the third plow in the floodplain environment of central Oukwanyama were disappointing, which was attributed to the drought conditions that had prevailed there during that year.[41] While these three plows once again had been acquired from a firm in South Africa via the good offices of the colonial administration (an elaborate and slow process), in the late 1940s the missions and at least one of the two wholesale shops in Ovamboland offered plows.[42]

The Assistant Native Commissioner at Oshikango, who was in charge of the late 1930s experiments, identified a number of obstacles to the introduction of the plow. He cited a shortage of cash to purchase plows and plowing equipment; cash could only be earned though migrant labor. A second challenge was a shortage of draft animals. Thirdly, he emphasized that cultivation was considered to be the work of women, but cattle, which were required to pull the plows, were the domain of men. Fourthly, he believed the inhabitants of Ovamboland to be ultraconservative, pointing to "[t]he inherent objection of Natives to adopt new ideas and abandon old methods". Lastly, he listed a lack of knowledge about how to use plows.[43] Nevertheless, by 1943, teams of trained draft oxen were widely enough available amongst Ovamboland's elite of chiefs and headmen to make it possible to employ a light animal-drawn grader to maintain the principal dirt roads across Ovamboland.[44]

In 1946, Ovamboland as a whole contained an estimated 100 plows although four years later, the total number of plows remained unchanged. Still, the number of trained draft animals, which would have been used to operate the plows, seems to have increased slowly, as is also demonstrated by the increase in the use of animal-drawn means of transport. Before World War II, very little mention is

[41] NAN, NAO 11, O/C Native Affairs Oshikango to NCO, Oshikango, 19 Sep. 1939, and NAO 20, Monthly Reports Ovamboland, Jan. – Feb. 1939.

[42] On the missions, see NAN, NAO 64, NCO to CNC, Ondangwa, 30 June 1948, and ANC, report Barter System of Trade in Native Areas, Oshikango, 28 June 1948. On purchasing plows at Ondjondjo Wholesale, Lea Paulus, interview by author, Onandjaba, 17 June 1993.

[43] NAN, NAO 11, O/C Native Affairs Oshikango to NCO, Oshikango, 19 Sep. 1939.

[44] NAN, NAO 25, NCO to CNC, Ondangwa, 16 Feb. 1943.

made of animal-drawn four-wheeled wagons and two-wheeled carts except for those that were used by the missions and the administration. The 1945–1946 Agricultural Census reported 50 wagons and 200 carts in use in Ovamboland, while the 1949–1950 Agricultural Census listed 75 wagons and only 50 carts.[45] It is possible that the census included mission- and government-owned wagons and carts, but the majority if not all of the vehicles probably were owned by Africans because the census was conducted by the Department of Native Affairs.

At the end of 1952, however, the number of plows had increased tenfold, to 1,073. The availability of trained draft animals again is reflected in the number of wagons and carts in use in 1952: 19 wagons and 576 carts. The Oukwanyama district accounted for over half of the plows and a quarter of the carts, and Ombalantu and Uukwambi each had approximately one-fifth of the total number of plows while only 39 were in use in Ondonga district. But Ondonga had 11 wagons (more than half of the total of 19), and 100 of the 576 carts, and it was the only district that had motorcars (there were 6 of them). Again, it is not entirely clear if the numbers include mission and administration property, but King Martin of Ondonga owned at least one motorcar in the early 1940s.[46] By 1952, the use of wagons and carts had become an annoyance to colonial officials because their metal wheels and the animals' hooves spoiled the surfaces of the sand roads, making them difficult for the motorcars of colonial officials to drive on. The official at Oshikango had requested the Oukwanyama headmen to keep wagons and carts with iron wheels off "our roads".[47]

Colonial officials considered the small number of plows in Ondonga to be evidence of Ovambo resistance to modernization, but the large number of wagons and carts in Ondonga, which required trained draft animals, and the presence of motorcars suggest that this cannot be the full explanation.

By 1957, according to the agricultural officer for Ovamboland, the plow was used on 20 % of the total arable area:

---

45 NAN, NAO 103, Censuses of Agriculture Ovamboland 1945–1946 and 1949–1950.

46 NAN, NAO 103, ANC to NCO, Oshikango, 30 Dec. 1952; Chief Kambonde to NCO, Okaroko, 18 Dec. 1953; Council of Headmen Ombalantu to NCO, 25 July 1952; Council of Headman Ukuambi to NCO, Ukuambi, 16 July 1952; Chief Ushona Shimi to NCO, Okakua, 7 July 1952; and Ikasha Nkandi and Ashimbanga Mupole to NCO, Onkolonkathi, 26 June 1952.

47 NAN, NAO 62, NCO to ANC, 19 Sep. 1952 and ANC to NCO, 22 Oct. 1952.

> The reason that the Ovambos increasingly make use of the plough, however, is that it requires much less labour and time to cultivate a field with a plough than to raise cultivation beds in the same field. Because the plough is also much faster every plough owner can take care of a bigger plot (...) Therefore although ploughed land produces less per field, the total production is higher because the farm plot can be increased.[48]

The administration actively promoted the use of the plow, for example by renting out tractors and encouraging farmers to purchase them. In 1976, renting a tractor with a driver cost ten rand per hour. In 1980, an estimated 100 tractors were in private hands and 20,000 ha were cultivated with plows. In the early 1990s, the large majority of households surveyed relied on plow cultivation.[49] In promoting the adoption of plows, however, the administration ignored reports that plowing, especially deep mechanical plowing, was detrimental to soil fertility. Based on 1970s trials in Ovamboland, the territory's Secretary for Agriculture informed the Secretary for Bantu Affairs in Pretoria that tractor (deep) plowing could cause the saline subsoil of Ovamboland to be mixed in with the thin topsoil, which depressed crop yields.[50]

In addition, the dissemination of the animal-drawn plow after the 1940s contributed to arable land scarcity because it facilitated and required the cultivation of larger fields per household, at the expense of farm and village pasture and bush lands. The labor-saving plow enabled larger fields to be prepared but, in combination with an increase in male absence due to migrant labor beyond Ovamboland, the use of the plow compounded the weeding bottleneck because weeding continued to be done by hoe. Weed competition decreased yields per hectare and consequently forced households to increase the area under plow cultivation, again at the expense of bush and grazing land.[51]

---

48 NAN, BAC 133, Agricultural Report Ovamboland 1956–1957.

49 NAN, OVA 50, Sec. Agriculture to Sec. Bantu Administration, Ondangwa, 2 April 1976, and J. Amutenya to Sec. Agriculture, Ombalantu, 30 Aug. 1975 and 13 Oct. 1976; OVA 6, Annual Report Agriculture Ovamboland 1979–1980; WW A 637, report appended to Erasmus to Director Water Affairs, Otjiwarongo, 13 May 1970; OMITI A5.2.2.

50 NAN, OVA 50, Sec. Agriculture to Sec. Bantu Administration, Ondangwa, 2 April 1976, and OVA 47 f. 6/8/3/1–7, Venn, Loxton & Associates, Mahanene Research Station Visit by Research Committee, 23–24 Feb. 1976. See also WW A 644, A. Trevor, ACE Planning, 11 July 1972; WW A 640, Report Ovamboland Pipelines, Oct. 1977; OVA 49, Meeting Subcommittee Townplanning, 2 Sep. 1970; OVA 93, Sec. Agriculture, 13 Sep. 1979.

51 Kreike, *Re-creating Eden*, chap. 6.

The introduction of plow agriculture also had social repercussions that affected the gender division of labor, female control over land and crops, and agricultural productivity. In 1993, only 37 % of a sample of 54 women had a field of their own, although 59 % emphasized that they had had their own field before they had married.[52] The loss of access to the proceeds from their own fields meant that women, who increasingly had become the mainstay of agricultural labor because of male absence due to migrant labor, had less incentive to invest additional labor in cultivation, for example, to do the extra weeding that plowing required. As a result, to some extent, even adult female labor may have been disinvested from crop cultivation from the 1950s onward.[53] Thus, in the social and environmental context of Ovamboland, the introduction of the plow had contradictory repercussions related to the intensification of crop cultivation. On the one hand, it led to an *intensification* of agriculture, with the adoption of ox- or donkey-drawn plows. On the other hand, crop cultivation became more labor- and land-*extensive*.[54]

Plowing also directly and indirectly affected the use and the availability of woody resources. The impact was direct because trees and tree stumps hindered plowing and it became more common to burn out tree stumps, especially when tractors were used, which, by 1993, was the case for 34 % of OMITI survey respondents.[55] Moreover, the plow meant that saplings were more easily plowed under and the root systems of existing trees were damaged. Still, that the plowshares cut the roots in some cases actually may have encouraged vegetative regeneration, as occurred, for example, with the marula tree (*Sclerocarya birrea*), because new trees developed from the cut roots.[56]

---

[52] Kreike, *Re-creating Eden*, chap. 6; OMITI A0.11 and 12 (N = 54).

[53] See, for example, Kreike, *Re-creating Eden*, chap. 5.

[54] Berry emphasizes that agricultural intensification is not a necessary result of, for example, population pressure and also notes that it is not irreversible, Berry, *No Condition Is Permanent*, pp. 181–196. Gray notes that despite population pressure, agricultural change in southwestern Burkina Faso in modern times has been marked by agricultural extensification, Gray, "Investing in Soil Quality".

[55] Lea Paulus, interview with author, Onandjaba, 17 June 1993, and NAN, NAO 26, Report Ovamboland Cotton Prospects, appendix to Alec Crosby to Bishop of Damaraland, St. Mary's Mission, 11 Jan. 1924. In the early 1960s, trees were common in fields in the eastern side of the middle floodplain and the area directly to its east, BAC 131, Deputy Secretary of Forestry, "Report of a Visit by the Deputy Secretary of Forestry (...) 17–29 April 1961", Pretoria, 10 May 1961. On the use of tractors, see OMITI A5.2.2. Tractor plowing greatly increased the possibility that tree trunks and roots would damage a plow blade, personal observations by author, 1991–1993.

[56] Interviews by author: Lea Paulus, Onandjaba, 17 June 1993; Helemiah Hamutenya, Omuulu Weembaxu, 17 July 1993; Philippus Haidima, Odibo, 9 Dec. 1992, and Pauline Onenghali, 15 Dec.

Indirectly, plowing and the entire social and agricultural complex within which the use of the plow became embedded affected the on-farm and off-farm availability of woody vegetation in the villages. Off-farm, the expansion of arable land as a result of an increasing number of farms per village and / or the expansion of individual farm plots diminished the total surface of the commons that was under woody vegetation. A diminishing village commons reduced the local availability of forest products and forage, and the scarcity of the latter in turn forced cattle owners and herdsmen to herd the cattle to the cattle posts earlier and for longer periods of time, reducing the availability of manure and other important cattle-derived products.

The agricultural report for 1955–1956 noted that the number of farm plots was increasing in all villages, and explained "It is not rare to see a native who cuts out hundreds of mopane trees in the mopane forests and then just leaves the trees to rot while he does not even plant manna [millet] on the clearing area". The author of the report feared that the consequences might be environmentally disastrous in the long term. He estimated that each new farm diminished a village's pastureland by 1.7 ha and noted that leaving the trees to rot destroyed years worth of potential firewood. Moreover, "[b]ecause the kraal and the field is kept clean there is no possibility that the area in the future will produce new trees that could be used as firewood (...). The presence of the new kraal also means that there is an additional consumer of firewood in the ward". Finally, he expressed concern that the remaining pasturages would be overgrazed, preventing tree regeneration, and he predicted that without trees, soil erosion would become a serious menace.[57]

Although associated with the social and environmental impacts outlined above, the impact of the plow was not unambiguously negative. The introduction of the animal-powered plow to some extent compensated for the increased loss

---

1992; Kreike, *Re-creating Eden*, chap. 6. NAN, NAO 62, Agricultural Report Ovamboland, 30 Nov. 1953; BAC 132, Agricultural Officer Ovamboland to NCO, Ondangwa, 1 March 1957; BAC 133, Agricultural Report Ovamboland, 1956–1957; WW A 637, report appended to Erasmus to Director Water Affairs, Otjiwarongo, 13 May 1970; OMITI A5.2.2. On the weeding bottleneck, see Richards, *Indigenous Agricultural Revolution*, p. 136.

[57] NAN, BAC 133, Agricultural Report Ovamboland 1955–1956. See also interviews by author: Joseph Kambangula, Omboloka, 25 Feb. 1993; Nahango Hailonga, Onamahoka, 4 Feb. 1993; Timotheus Nakale, Big Ekoka, 21 Feb. 1993; and Moses Kakoto, Okongo, 17 Feb. 1993.

of male labor to agriculture as a result of the growing male employment outside of Ovamboland. Before World War II, men and boys from the district of Oukwanyama—the largest supplier of migrant labor to the white farms and the mines of Namibia—chose to engage in migrant labor away from home, typically after the major household agricultural labor needs had been fulfilled. The only exception was during years of drought, which destroyed the prospects for a good harvest. During and after World War II, household agricultural production became secondary for men and older boys: the animal-powered plow made it possible to divert male labor from household agriculture on a massive scale while simultaneously maintaining overall crop production by putting more land into production.[58]

Animal-drawn plows also facilitated the cultivation of land that had been unsuitable for hoe cultivation. Before the adoption of the plow, the lighter sandy soils on the middle slopes of the 'dunes' between the seasonal water courses were the preferred location for fields because the heavier clayish soils were too difficult to cultivate with the hoe. The animal-powered steel plow, however, dramatically increased the land available for crop cultivation in the floodplain because land closer to water courses and even within the upper flood levels could be cultivated. Outside of the floodplain, the plow also made it possible to cultivate the heavy soils on the edge of and within pans and other shallow seasonal bodies of water.[59] The availability of additional land relieved the pressure on environmental resources in the densely settled floodplain as people migrated east and beyond the floodplain, resulting in a more even distribution of the area's human and livestock populations.

Still, why is it that plow use was insignificant for decades after plows were introduced in World War I, and what caused the technology to be rapidly adopted after World War II? One limitation certainly was availability. Plows were expensive: in 1918, a plow cost seven pounds sterling while monthly migrant labor wages at

[58] On migrant labor and the plow as a labor-saving device, see Kreike, *Re-creating Eden*, pp. 81–99.

[59] Erastus Shilongo, interview by author, Olupandu, 21 June 1993; Abisai Dula, interview by author, Oikokola, 25 June 1993; NAN, NAO 101, Senior Agricultural Officer Natives, Agricultural Survey of Ovamboland with Reference to Agricultural and Stock Improvement in that Area, Windhoek, 26 Oct. 1947. See also Kreike, *Re-creating Eden*, pp. 152–153.

the Namibian diamond fields were about one pound a month.[60] Moreover, outlets for acquiring plows were limited before the 1940s. The colonial state imported few plows and the missions only sold plows in Ovamboland from the late 1940s onwards.[61] Until 1939, only three stores operated in Ovamboland, but between 1939 and 1947, their price lists do not reference any plows or plowing gear. The inhabitants of the Oukwanyama district also had access to traders from across the border and they also purchased considerable quantities of goods in Angola, but there is no evidence of any substantial cross-border trade in plows.[62]

Plows became more prevalent during the late 1940s and the 1950s; larger numbers of men and boys engaged in wage labor and wages increased while the prices of imported steel and iron implements and tools fell. From 1947 onwards, plows were included in the price lists of goods sold at the OTC stores; in 1947, the price was £5 7s. 6d. In 1938, the annual wage for the highest class of workers recruited for the Consolidated Diamond mines was £14 8s. The lower-paid workers received £5 8s.[63] Wages were still around one pound a month after the war, although laborers who developed special skills were paid substantially more; for example, in the late 1940s, Julius Abraham made five pounds a month as a railroad worker.[64] During and after World War II, more men and older boys engaged in migrant labor for increasingly longer periods and they did so more frequently, which provided individuals and households with more money to invest in plows and draft animals.[65] In the 1940s, when she was an adolescent, Lea Paulus plowed her uncle's fields at Okalongo with an ox plow. Her uncle was a schoolteacher with a good salary who was one of the first in the area to purchase a plow, which he bought at the OTC's Ondjondjo shop at Ondangwa in the 1940s.[66]

---

60 Kreike, *Re-creating Eden*, p. 89: Petrus Shanika Hipetwa, Oshiteyatemo, 17 June 1993. 1949: 1s. 6d/day, NAN, NAO 89, NCO to CNC, Windhoek, 22 July 1949.

61 On the missionaries selling plows, see NAN, NAO 64, NCO to CNC, Ondangwa, 30 June 1948; ANC, report on Barter System of Trade in Native Areas, Oshikango, 28 June 1948.

62 NAN, NAO 23, NCO to Sec. SWA, Ondangwa, 1 Aug. 1936.

63 Kreike, *Re-creating Eden*, p. 217.

64 On wages, see interviews by author: Paulus Wanakashimba, Odimbo, 10 Feb. 1993; Erastus Shilongo, Olupanda, 21 June 1993; Joseph Nghudika, Onamahoka, 3 Feb. 1993; Julius Abraham, Olupito, 16 June 1993. On higher wages on the South African Rand, see NAN, A450, 13, SWA Native Labour Commission 1945–1948, Minutes of Meetings and Testimony 1947, vol. 2: Claus C.S. Holdt, ANC, Oshikango, 1 Sep. 1947.

65 Kreike, *Re-creating Eden*, pp. 90–99.

66 Lea Paulus, interview by author, Onandjaba, 17 June 1993.

Not only did the supply of plows increase and the price drop, but, in addition, more trained or trainable draft animals became available. Interviewees stress that although oxen were used for plowing, donkeys were a much more affordable alternative for plowing and for transport. Plowing immediately followed the first good rains, before village and farm sources of water and grazing had recovered. In years of drought, however, oxen and donkeys were often too weak to plow.[67] But donkeys were more expendable than oxen. Moreover, unlike cattle, donkeys were not part of the dry season transhumance trek to the distant cattle posts; they were kept near the villages and thus immediately were available. To use oxen for plowing was a high-risk undertaking because the cattle posts could be located one to two weeks' distance from the villages. After World War II, migrant laborers imported large numbers of donkeys.[68] The supply of donkeys thus may have been as much a precondition for the adoption of the plow as a consequence of it.

Another inhibiting factor was that the conventional use of the plow was incompatible with the creation and use of arable land. When a plot was cleared for a field, floodplain farmers allowed the stumps and the roots of large trees to remain behind. The Anglican agricultural missionary Alec Crosby in 1924 despaired over the "uneconomical" and inefficient way floodplain farmers cleared the land: "The usual method is to put a fire around a tree until it falls, no effort being made to remove the stump". He railed against how the method destroyed natural vegetation, but the alternative that he proposed can hardly be considered more environmentally sensitive: "Before ploughing can be really effective it will be necessary to stump the ground either by blasting or by digging the roots out".[69] When she was an older girl during the 1940s, Lea Paulus "worked like a boy because I mounted the 4 oxen on the plow and plowed the entire field like the young men (...). It was difficult to plow because if you hit a tree stump

---

[67] Interviews by author: Erastus Shilongo, Olupandu, 21 June 1993; Kaulipondwa Tuyenikalao Augustaf, Odimbo, 10 Feb. 1993; Hendrik Hamunime, Eko, 21 May 1993; Salome Tushimbeni, Oipya, 19 June 1993. NAN, NAO 100, statement of Ruusa Niinkoti Mangundu, appendix to NCO to Chief Kambonde, Ondangwa, 24 Nov. 1954; BAC 44, Minutes Annual Meeting Ondonga, 1 Dec. 1958, and BAC 45, Minutes Tribal Meetings Oukwanyama, 9–29 May 1958.

[68] See chapter 4.

[69] NAN, NAO 26, Report Ovamboland Cotton Prospects, appendix to Alec Crosby to Bishop of Damaraland, St. Mary's Mission, 11 Jan. 1924.

the handles of the plow could hurt you. When the plow was introduced people started to cut down trees and the remainder of the trees was set on fire".[70]

During the late 1930s and 1940s, floodplain farmers invented a variety of strategies to combine using the plow with a measure of on-field flood management. Ovambo farmers grew their crops on raised beds for several reasons; one reason was to drain surplus rainwater away from the fields through the 'paths' in between the crop beds. Farmers constructed the beds with their hoes after the first good rains had fallen. The 1930s experiments with plows conducted by, for example, Senior Headman Eliah Weyulu and Lea Paulus' uncle demonstrated the danger of abandoning the raised bed system since it made the crops vulnerable to flooding.[71] After World War II, farmers subsequently chose to retain drainage paths across their plowed fields, either by crisscrossing their fields in checkerboard fashion, effectively imitating the raised bed pattern that marked pre-plow cultivation, or by plowing drainage 'paths' along the contour lines or down the slope to the seasonal flood channels.[72]

Social dynamics relating to Christianity, class and gender also converged to bring about the adoption of the plow during the 1940s and 1950s, at the same time that the plow brought about critical social changes. During the 1930s, the elite of headmen and clan elders clashed with the nascent Christian elite of pastors and teachers. Missionaries and their Christian followers contested the clan elders' and the headmen's control over property (in particular land, cattle, water and trees) and people through the institutions of female initiation and marriage. Christians refused to have Christian girls participate in the initiation ceremony and they rejected the notion of paying a bride price, which they considered to be slavery. They also rejected the authority of a married woman's clan elders over the woman's person, her property and her children; typically the woman and her children all remained members of her matrilineal clan with full rights to her property invested in her clan. With the colonial administration strongly supporting the headmen, the Christian elite initially had to back off. During the late 1940s and 1950s,

[70] Lea Paulus, interview by author, Onandjaba, 17 June 1993.

[71] NAN, NAO 11, O/C Native Affairs to NCO, Oshikango, 19 Sep. 1939.

[72] Lea Paulus, interview by author, Onandjaba, 17 June 1993; Erastus Shilongo, interview by author, Olupandu, 21 June 1993; NAN, NAO 62, Agricultural Report Ovamboland, Omafu, 30 Nov. 1953.

however, conflict began anew, only now the administration and many headmen as well as many married men were allied with the Christian elite. By the early 1960s, opposition by male heads of households had weakened the grip of their wives' clan elders over their wives and children and their household property. One of the results was that the identification of cattle as an exclusively male domain was strengthened and, combined with the plow, which could be bought with wages that only men could earn, enabled men to make the 'ox plow' an avatar to secure their dominance in a rural agricultural space that they had all but abandoned during the 1940s and the 1950s, when they were so heavily engaged in migrant labor.[73]

Guns and steel in north-central Namibia

The environmental impact of Western gun and steel technology has been ambiguous. The proliferation of firearms in the Ovambo floodplain prevented European conquest until World War I. Guns made hunting more effective, especially elephant hunting to obtain ivory which in turn bought more firearms and horses. The introduction of firearms shaped violence in the floodplain beginning in the second half of the nineteenth century, and their use increased the scope and intensity of cattle raiding, especially after rinderpest decimated the herds. Disarmament in the early colonial era allowed a recovery of game populations in the region, but also contributed to the maintenance of elaborately palisaded homesteads along with fenced fields as a protection against wild animals, and thus contributed to deforestation.

Although Western industrially produced tools were available since before World War I, blacksmiths in the Ovambo floodplain held their own with far superior steel hoes until after World War II. Similarly, the adoption of other Western steel technology was not linear and predetermined. The use of the animal-drawn metal plow has ancient origins in the Ethiopian highlands, but elsewhere in Africa the plow was not introduced until early in the colonial era. The same was true in north-central Namibia, where the animal plow was first introduced before World

[73] Kreike, *Re-creating Eden*, pp. 81–128.

War I, but was not widely adopted until after World War II. Moreover, the donkey, not the conventional ox, emerged as the main plow animal in north-central Namibia with repercussions for land use. The animal-drawn plow caused yet further deforestation and contributed to increased land scarcity, but the technology also relieved environmental pressures in the floodplain because it facilitated the settlement of Ovamboland east of the floodplain. Socially, increased dependence on imported metal tools such as the plow required more of the male population to hire themselves out for wages outside the region, while the hoe continued to be used by women to weed ever-growing plots. The use of both implements affected agricultural production, resource management and social relations, with women becoming the de facto environmental resource managers, while men struggled mightily to retain de jure control.

6

Naturalizing cattle culture: Colonialism as a deglobalizing and decommodifying force

Models of environmental change derived from the Nature-Culture dichotomy posit a precolonial, undifferentiated Natural subsistence economy that is penetrated by a colonial, market economy Culture. In the modernization paradigm of environmental change, the interaction is seen as positive: natural resources are held to be more effectively used in a market economy. In the declinist and inclinist paradigms, the result is perceived as environmental degradation. The history of cattle management and use in the Angolan-Namibian border region between 1890 and 1990, however, once again complicates unilinear Nature-to-Culture narratives of environmental change.

An influential 1980s report by Keith Morrow, an agricultural expert, stated:

> Despite the limited potential within Owambo [Ovamboland] the natural resources are being used at a fraction of their potential and much of this involves the misuse and deterioration of natural grazing. Livestock production practices are primitive because of a lack of knowledge and training and the absence of an acceptable market, thus minimising the annual offtake (...). [I]n the

> absence of any acceptable market outlets it is not possible to persuade cattle owners to adopt accepted commercial practices of animal husbandry.[1]

This statement relegates human livestock management and use in north-central Namibia to the category of 'primitive' and hails modern scientific knowledge and the market in tandem as the solution to environmental degradation. Human mismanagement of the environment is a major concern of the modernization, declinist and inclinist paradigms of environmental change. The declinist view is that modernity (where Culture dominates and exploits Nature) engenders destructive environmental behavior through a global market-driven overexploitation of the local resource base, and / or a cattle population explosion that results from the introduction of veterinary science that reduces cattle disease mortality. The modernization paradigm identifies tradition as the culprit: primitive and irrational pastoralist practices lead to an inefficient and wasteful use of resources, causing, for example, overgrazing. The inclinist paradigm, by contrast, celebrates tradition, based on the premise that only indigenous, small-scale cattle management and use that mimics or approximates a 'state of Nature' is sustainable.[2]

Cattle actually were a global market commodity in the Ovambo floodplain before the colonial conquest; only during and because of colonial rule were cattle reduced to a resource for local trade and subsistence. The alarm about overgrazing, deforestation and desertification was raised by colonial officials and experts who by their own admission were unwilling and unable to 'modernize' the cattle sector. Yet, there is little evidence to support their claims of serious environmental degradation. On the other hand, the record does not support the assertion that traditional indigenous management and cattle use in north-central Namibia were stable and naturally sustainable either. Rather, pastoralism in the region was subject to dramatic upheavals caused by war, disease and migration.

The case of Ovamboland does not unambiguously support the thesis that a livestock 'population explosion' resulted in overstocking. Rather, the region's livestock population was subject to radical fluctuations, with overall numbers

---

1 [Keith Morrow], "A Framework for the Long Term Development of Agriculture within Owambo", Aug. 1989.

2 On management see Blaikie and Brookfield, *Land Degradation and Society*, pp. 3, 27–48, 100–156; and Gibson, McKean and Ostrom, *People and Forests*, pp. 1–85, 135–161, 193–226.

declining in the late 1890s, the 1910s and the 1980s, defying linear models of environmental change. In the 1920s and 1930s, cattle became viewed as a health threat that had to be strictly quarantined. By the late 1940s, Ovamboland's cattle were considered to be a major environmental obstacle to development, a view that persisted well beyond the late 1980s.

The cattle complex and environmental degradation

The idea that African pastoralists did not consider cattle to be commodities, but rather cultural objects, for example, as signs of wealth, status, prestige or piety, was prominent in explanations of overstocking in colonial Africa in the 1960s. Herskovits coined the expression 'cattle complex' to describe this phenomenon.[3] As it relates to environmental degradation, the cattle complex argument explained that because of the animals' high cultural value, the management objective was to maximize their numbers by minimizing their consumption and sale. The practice of hoarding cattle resulted in a cattle population explosion and overgrazing and desertification. The theory suggested that if livestock managers were to behave 'rationally', that is, respond to market opportunities and employ modern cattle management practices, the cycle of overstocking-degradation would be broken because 'surplus' cattle could be sold and consumed.[4] Africa's cattle complex, however, may be little more than a recent (re)invention of tradition, and indeed the case of Ovamboland suggests that it was such a recent invention that it could hardly be called a tradition at all.[5] In that context it is significant that colonial officials and experts emphasized their view that cattle management in Ovamboland was 'primitive', but they did not employ the cattle complex theory in Ovamboland until the 1980s.

---

[3] Herskovits, "The Cattle Complex in East Africa".

[4] See Scoones, "Range Management Science and Policy", and Beinart, "Soil Erosion, Animals, and Pasture over the Longer Term". See also Swift, Conghenour and Atsedu, "Arid and Semi-Arid Ecosystems". On livestock overpopulation, see Le Houérou, *The Grazing Land Ecosystems of the African Sahel*, pp. 90–128; and Beinroth, "Land Resources for Forage in the Tropics". For India, see Jha, *The Myth of the Holy Cow*.

[5] Ferguson, *The Anti-Politics Machine*; and Cohen and Atieno Odhiambo, *Siaya*, p. 76.

As elsewhere in Africa, reports for Ovamboland demonstrate that the cattle population increased rapidly during the colonial era, but time series figures for domestic animal population on the continent are often crude estimates. An influential study of cattle raising in the Sahel countries of West Africa, for example, uses livestock figures from 1950 and 1983 to demonstrate that the numbers more than doubled. If the available figures for 1968 and 1973 are added, however, the trends become less linear. In fact, the cattle population of Burkina Faso actually declined between 1968 and 1973 and the 1983 figures did not dramatically surpass the 1968 level. In South Africa between the early 1970s and the late 1980s, numbers of small stock were on the whole lower than at any time since the first decade of the century and overall the numbers were relatively stable over time.[6]

Moreover, assessing the impact of domestic animals on the environment as such is a challenge. The key scientific concept of carrying capacity—the number of animals an environment can sustain without structural degradation—is highly contested.[7] An increase in the ratio of unpalatable species in pasturage—either poisonous or woody plants (the spread of the latter being referred to in southern Africa as bush encroachment)—frequently is used as an indicator of environmental degradation.[8] Poisonous plants, however, are often a "natural part of high condition range communities". They cannot be defined as 'bad' or interpreted as an indication of degradation simply because they happen to be poisonous to livestock.[9] Moreover, some 'poisonous' plants are at the same time important dry season sources of animal browse, including sorghum in Africa, oak in Europe and North America, and the one-time global agroforestry 'miracle tree', the native Latin American lead tree (*Leucaena leucocephala*). Indeed, oak and lead tree are among many plants that contain chemicals that have poisonous effects if they are digested in large quantities or if they form the bulk of a livestock diet. Careful live-

---

[6] See Le Houérou, *The Grazing Land Ecosystems of the African Sahel*, pp. 124–126, tables 24–28, and Beinart, "Soil Erosion, Animals, and Pasture over the Longer Term", p. 66.

[7] On carrying capacity, see Scoones, "Range Management Science and Policy", and Beinart, "Soil Erosion, Animals, and Pasture over the Longer Term"; Little, "Rethinking Interdisciplinary Paradigms and the Political Ecology of Pastoralism in East Africa", pp. 163–164; and Munro, "Ecological 'Crisis' and Resource Management Policy in Zimbabwe's Communal Lands", p. 195. See also Simon, "Sustainable Development".

[8] Le Houérou, *The Grazing Land Ecosystems of the African Sahel*, pp. 90–128.

[9] Laycock, Young and Uechert, "Ecological Status of Poisonous Plants on Rangelands", and Ralphs and Sharp, "Management to Reduce Livestock Loss from Poisonous Plants".

stock management is the key to preventing overfeeding on any of these plants, and increased incidences of cattle poisoning may be the result of a deterioration of cattle management rather than an indicator of an increase in the ratio of poisonous plants in pasturage and the deterioration of grazing.[10]

The sections that follow analyze the impact of cattle management and use in Ovamboland within the context of Nature-to-Culture narratives of environmental change.

Ovambo cattle as global commodities

Before the colonial era and on the eve of the 1896–1897 rinderpest epidemic that decimated cattle herds, the inhabitants of the Angolan-Namibian border region bred and exported large numbers of cattle. The semi-arid region was free of tsetse fly, the vector for sleeping sickness in humans and *nagana* in cattle that severely limits pastoralism across Africa.[11] Cattle were kept close to the floodplain villages during the rainy season. To conserve precious water and forage resources during the long June to December dry season, herdsmen drove the cattle to distant cattle posts outside of the Ovambo floodplain until the return of the rains. The dry season cattle post areas in the 1890s included the Kaokoveld (Kaokoland) to the west of the floodplain, Etosha Pan to the south and Oshimolo (in Angola) to the northeast.[12]

Cattle were the major export from the Ovambo floodplain in the late 1870s through the early 1900s. The kings of Oukwanyama, the largest of the Ovambo floodplain's polities, were the most prominent suppliers of cattle to European traders, who shipped the animals to the Cape Colony and the Transvaal as well as to Luanda, St. Helena and Gabon. The Oukwanyama kings alone may have

---

[10] On oak, see Ralphs and Sharp, "Management to Reduce Livestock Loss", and Harper, Ruyle and Rittenhouse, "Toxicity Problems Associated with the Grazing of Oak in Intermountain and Southwestern USA". On sorghum, see Hanna and Torres-Cardona, "*Pennisetums* and *Sorghums* in an Integrated Feeding System in the Tropics", esp. pp. 195–196. On *Leucaena*, Lawton, "Browse in Miombo Woodland", p. 30; and Shelton and Brewbaker, "*Leucaena leucocephala*". In general, see Huxley, *Tropical Agroforestry*, pp. 39–50.

[11] Kreike, *Re-creating Eden*, pp. 21–25.

[12] Kreike, "Architects of Nature", pp. 107–108.

supplied an annual average of 600 head of cattle in the decade or so before the 1897 rinderpest and an average of 2,000 head annually in the decade or so following the disease, even though the epizootic dramatically decimated cattle herds.[13]

Human settlement and movement had a critical impact on animal use and management while war and violence dramatically affected settlement patterns during the early decades of the twentieth century. Portuguese colonial conquest and pacification followed in the wake of the rinderpest, causing massive death and destruction in the northern Ovambo floodplain as well as further decimating livestock herds. Cattle meat was critical for the survival of the refugees from the northern floodplain, who found shelter from the violence of war in the uninhabited wilderness areas in the middle Ovambo floodplain around the modern Angolan-Namibian border.[14]

Cattle, culture and nature

Despite the rinderpest, Oukwanyama's kings and princes more than tripled the number of cattle they supplied to the global market in the early 1900s. Higher cattle prices and an arms race spurred by the threat of European expansion explain the higher level of cattle commodification. The Oukwanyama kings and princes obtained much of the export cattle through violent raids against neighboring polities (whose elites responded in kind) and against their own subjects.[15] This process mirrors the impact of global markets on indigenous *wild* animal use in general, including, for example, the African elephant and the American bison, causing overexploitation, heightened strife over a dwindling precious resource and environmental degradation.[16]

Cattle were not merely a global export commodity in the early 1900s; they were also critical local material and social currencies. As the cattle complex theory predicts, cattle were a major source of wealth, indeed wealth was measured in cattle and the local millet staple. In his ethnographic manuscripts, Native

---

13 Kreike, *Re-creating Eden*, chaps. 2–3.
14 Ibid., chap. 4.
15 Ibid., chaps. 2–3.
16 MacKenzie, *The Empire of Nature*, and Isenberg, *The Destruction of the Bison.*

Commissioner C.H.L. Hahn wrote: "[A man's] richness are [*sic*] generally gaged [*sic*] by the (...) [?] size of his granaries and the extent of his herds". Headman Shimwefeleni was considered one of the wealthiest men in Ovamboland because he owned large numbers of livestock and extensive stores of millet.[17] Annual cattle fests that marked the return of the herds to the villages were occasions to acknowledge the expertise of the herdsmen and to display the wealth of the cattle owners, in addition to serving as public accounting exercises. In May 1916, a colonial officer witnessed the cattle fest that assembled the Oukwanyama King Mandume's impressive herd. A painful discovery made during Ondonga Queen Mtwaleni's well-attended cattle fest in July 1938 was that her husband had absconded with many of her cattle.[18]

But cattle were not only hoarded and bred for display. Cattle were also redistributed to enhance and maintain social networks as well as to spread risk by not concentrating all a person's cattle in one pen. In keeping with the moral economy thesis, faced with the ever-present threats of drought and disease, people maximized investment in social networks (household, family, clan, village, patrons) that they tapped in times of need.[19] Rather than simply hoarding or selling 'surplus' cattle, owners used the animals to cement relationships through feasting, gifting (bridewealth) and lending arrangements.[20]

The moral economy framework is compatible with the cattle complex model because it emphasizes local culture and initiative alongside resistance to marketing cattle. But the moral economy approach stresses the social roots of the unwillingness to market cattle whereas the cattle complex thesis highlights vague and ahistorical cultural roots. The moral economy model enriches the cattle complex analysis because it shifts the emphasis from hoarding to circulation of cattle. The cattle complex thesis highlights maximizing the number of cattle in the hands of individuals; the moral economy model introduces nuances beyond sheer numbers: for example, cattle can be exploited beyond the actual number of animals because the increased circulation of cattle multiplies social relations within society.

---

[17] NAN, A450, 10, Family Life, and vol. 7, Annual Report Ovamboland 1938.

[18] NAN, UNG, UA 1, Fairlie, Information re the Property of the late Chief Mandume, Namakunde, 24 April 1917, and NAO 19, Monthly Reports Ovamboland, July – Aug. 1932.

[19] See Scott, *The Moral Economy of the Peasant*, and Spear, *Mountain Farmers*.

[20] Kreike, *Re-creating Eden*, pp. 101–128, 158–176.

The moral economy framework in and of itself does not explain, however, why before 1915, cattle in the Ovambo floodplain region were so readily sold to global markets, or why in the 1910s, 1920s and 1930s, northern floodplain Portuguese colonial subjects eagerly offered cattle for sale. In 1910, the Portuguese colonial official in charge of the recently occupied Ombadja polity in the northwestern Ovambo floodplain complained that his subjects too readily parted with their cattle to pay their taxes, thus evading the need to engage in wage labor to pay the taxes in cash. He recommended prohibiting paying colonial taxes with cattle. In 1914, the supply of cattle remained high: in Humbe, a major Portuguese commercial center in southern Angola, located just across the Kunene River from the Ovambo floodplain, local cattle owners went from trader to trader to sell their cattle. Because there was not enough money available to purchase all the cattle on offer, many of the cattle owners took their cattle 'elsewhere'.[21]

'Elsewhere' was down south in the Ovambo floodplain. After 1915, South Africa-occupied Ovamboland became a net importer of cattle and the principal market for cattle from the northern floodplain. Another source of import cattle was Namibia, south of Ovamboland and Etosha Pan, an area that was occupied by white settler ranchers and farmers and that was referred to as the Police Zone. From the 1920s, the Portuguese officials in south-central Angola only accepted tax payments in cash. Rather than selling their animals for artificially low prices to Portuguese traders, Angolan cattle owners sold their cattle to young men from south of the border who had earned money as migrant laborers in the mines in the Police Zone. South-central Angola's cattle holders had plenty of cattle to sell: a 1935 Portuguese colonial report estimated the cattle population in their Lower Kunene district at 1.2 million head, nine times the number in South African-controlled Ovamboland. The figure has some validity because the Portuguese registered the stock owned by their African subjects.[22]

---

[21] CNDIH, Avulsos, Caixa 739 "Huila" (1885–1929), 16, Governo do Distrito da Huila, o Governador, Relatório, Lubango, 29 Oct. 1910; and Codices No. 2339/347, cota 8-2-28, Serviços de Fazenda, Antonio Maria Meirdes e Vasconcelos, Inspecção ao Distrito da Huila, Mapas e Documentes 1913–1914, No. 50.

[22] For the Angolan cattle numbers, see "Okuwah-ah-kana mosi-oa-tunia (Um Esboço da Regiào Kalahariana e dos Territórios do Sul de Angola)", *Boletim da Sociedade de Geografia de Lisboa* 62(7–8) (July – Aug. 1944): 461–471. On cattle registration, see NAN, NAO 16, O/C NAO to Sec. SWA, Ondangwa, 15 Jan. 1927. For the 1935 Ovamboland number, see Kreike, "Architects of Nature", p. 112, table 6.1.

Like young men throughout early twentieth-century Africa, migrant laborers from Ovamboland invested part of their hard-earned wages in cattle. This was especially the case amongst laborers originating from the refugee communities that had established themselves in the middle floodplain wilderness along the Angolan-Namibian border. By the late 1930s, migrant laborers could purchase a head of cattle for one to two pounds sterling, the equivalent of one to two months' wages in the mines. The supply of northern floodplain cattle kept prices so low in Ovamboland's border district of Oukwanyama that its cattle owners complained about unfair competition. In 1934, two residents from Ovamboland ventured across the border and bought an ox and a cow in southern Angola for only five shillings each, less than half the going price in Ovamboland. In the 1940s, 1950s and 1960s, cattle continued to be a major purchase for returning migrant laborers.[23]

As a result of the continuous import as well as of natural increase, the estimated 60,000 cattle in 1925 had grown to the total of 386,000 cattle that were vaccinated in 1957.[24] Native Commissioner Hahn commented in 1924 on what he perceived to be the pastoralization of Ovamboland:

> In former years the Ovambos were regarded as an agricultural people but they are gradually becoming a pastoral people also. Their cattle herds are not only fast increasing but the standard of their small and interbred stock is gradually being improved through the Ovambos having received every facility and assistance from the Administration to introduce into their country a much bigger and better class of stock from the south. In recent years considerable herds have found their way to Ovamboland. It is not uncommon to hear of Native headmen owning herds numbering from one to two thousand head. The ordinary rank and file are too today possessed of stock and much better off than they have ever [been] before.[25]

While the inhabitants of the South African-occupied southern Ovambo floodplain were characterized as traditional agriculturalists who diversified into

---

23 Kreike, *Re-creating Eden*, pp. 64–65, 81–82, 89–98. For Angolan prices, see NAN, NAO 20, Monthly Reports Ovamboland, July – Aug. 1938, and NAO 16, NCO to Chefe de Posto Namakunde, Ondangwa, 16 April 1934.

24 Kreike, "Architects of Nature", p. 112, table 6.1.

25 NAN, NAO 18, Hahn, Notes on Ovamboland for the Administrator, Windhoek, 15 May 1924.

pastoralism, the inhabitants of the Portuguese-occupied northern floodplain were considered to be real pastoralists.[26]

Although Ovamboland's inhabitants rebuilt their cattle herds, they did not regain access to global markets. In an undated 1920s or 1930s document, Native Commissioner Hahn stated that the inhabitants of Ovamboland had sufficient land and cattle to be "practically self-supporting" but that they had no markets outside the Ovamboland reserve. As a consequence, Hahn argued, Ovamboland's cattle owners saw no need to improve their livestock-raising methods which had remained stagnant for the last 150 years.[27] In Hahn's estimation, in the absence of markets, cattle were at best a source of subsistence. Given Hahn's three decades as Native Commissioner and his prominence as an ethnographer and the leading contemporary authority on the Ovambo and Ovamboland,[28] it is no wonder that some of Hahn's pre-World War II sentiments were echoed as late as the late 1980s (including the statement quoted at the beginning of this chapter).

Although Hahn viewed livestock management in Ovamboland as primitive, he disagreed that the Native Reserve was overstocked in the 1920s and 1930s. He acknowledged that a certain degree of environmental deterioration was occurring, but he believed that any current or future grazing pressure in Ovamboland could easily be relieved by developing water resources in the uninhabited areas surrounding the floodplain. He maintained his optimism even as his Portuguese counterparts increasingly obstructed Ovamboland's inhabitants' access to Angola's dry season cattle pastures north of the border. Hahn identified the physical absence of a cattle market as a potential source for overgrazing rather than any cultural inhibitions against selling cattle, as would be the case in the cattle complex model.[29]

Whereas Hahn encouraged the import of Angolan and Namibian cattle into Ovamboland and turned a blind eye to the export of cattle and cattle products from Ovamboland to Angola, he kept colonial Namibia strictly off limits to cattle

---

[26] *Aperçue historique: Chronique des missions confieés à la Congrégation du Saint Esprit*, p. 285; and AGCSSp 485-A-III, "Cubango-Angola: Reprise de la Mission du Cuanyama [1923?]".

[27] NAN, A450, 10, "Agriculture".

[28] Kreike, *Re-creating Eden*, pp. 85, 102, 109, 118–120, and Hayes, "'Cocky' Hahn and the 'Black Venus'".

[29] NAN, A450, 12, South West Africa Commission, Minutes of Evidence, vol. 12, Ukualuthi, 13 Aug. 1935, Evidence Hahn, and vol. 10, Agriculture; Kreike, *Re-creating Eden*, pp. 129–176.

and cattle products from the Native Reserve. No cattle from Ovamboland were allowed to enter either Namibia's white settler areas to the south (the Police Zone), or the Native Reserves to the west (the Kaokoveld) from the late 1920s onward and to the east (the Okavango) from the late 1930s onward.[30]

Even as Hahn cited the prevalence of lungsickness as a pretext for prohibiting the export of cattle, he believed that Ovambo cattle largely were resistant to the disease and he therefore settled for isolating Ovamboland's cattle.[31] As has been noted above, however, in the early 1940s Hahn appears to have been changing his mind about the environmental impact of livestock, at least in private. In what appears to be a draft of his 1942 annual report for Ovamboland, he wrote that 'overstocking' was increasingly obvious.[32]

Overstocking and biological time bombs

Following Hahn's retirement in 1946, the idea that Ovamboland was overstocked quickly gained currency. Increased conflicts over access to water and grazing for cattle during the last years of Hahn's tenure added to concerns about the environmental impact of livestock, and 'overstocking' became a central theme in colonial reports during the 1950s.[33] A 1952 report concluded that the large Oukwanyama district of Ovamboland was overstocked, and the 1953 annual health report for Ovamboland noted that "the overgrazing of areas denuded of trees, will amongst other things interfere with the water supply and multiply the danger of fly and tick borne diseases".[34]

---

[30] NAN, NAO 18–20, Monthly Reports Ovamboland, Nov. 1926, Nov. – Dec. 1936 and May 1940; AGR 25, Senior Veterinary Surgeon to Sec. SWA, Windhoek, 13 Nov. 1941; NAO 11, O/C NAO to NCO, 19 Sep. 1939. On export to Angola, see A450, 10, "Agriculture", and NAO 58, Director of Agriculture to Secretary SWA, 24 Dec. 1946.

[31] NAN, NAO 18 and 20, Monthly Reports Ovamboland, Nov. 1926, May 1939, and Annual Report Ovamboland 1929.

[32] NAN, A450, 10, [Ms., 1942?].

[33] NAN, A450, 7, Annual Reports Ovamboland 1941 and 1943; BAC 133, Agricultural Officer to NCO, Report of travel to the northwestern part of Ovamboland, 20–22 June 1956, Ondangwa, 4 July 1956 and Agricultural Report Ovamboland, 1956–1957.

[34] NAN, NAO 60, Quarterly Report Ovamboland Oct. – Dec. 1952, and NAO 65, Annual Health Report Ovamboland 1953.

Ovambo cattle were not only depicted as a disease time bomb, but were also re-imagined as a neo-Malthusian overpopulation and environmental time bomb. Colonial statistics were marshaled to demonstrate that between 1925 and 1975, the cattle population had increased by a factor of nine, from 60,000 to 530,000 animals. The figures are guesstimates, although the 1960s and 1970s data are more reliable because they are based on the actual numbers of vaccinated animals. Overall, the figures suggest a steady increase of Ovamboland's cattle population from its 1915 low to the mid-1980s, followed by a sharp decline.[35]

Upon Hahn's retirement, a senior agricultural expert conducted a survey of the agricultural potential in Ovamboland. Unlike Hahn, the expert was totally unimpressed by Ovamboland's indigenous cattle: "The average Ovambo large stock are the poorest grown animals I have ever seen and are a fine example of what toll nature will exact for poor grazing, uncontrolled inbreeding, non-selection and internal parasitic infection". The scientist thought that Ovamboland's cattle breed was of such poor quality and size that it was impossible to improve it through crossbreeding with Afrikander bulls.[36]

Hahn's successor as Native Commissioner for Ovamboland, Harold Eedes, concurred but was determined to tear the reserve from the grip of the forces of Nature. He scorned the idea that Ovamboland's pastoralist and agricultural sector could be developed before livestock diseases were eradicated. In 1949, Eedes publicly dismissed a report with recommendations from a senior South African agricultural expert who had toured Ovamboland before it had even been published. Eedes commented: "As long as these Native areas remain 'closed areas' with several cattle diseases such as Lungsickness, etc. etc., it does not seem to me that any useful purpose will be served by attempting the improvement of native-owned stock".[37]

Eedes had led a 1938 inoculation campaign that successfully eradicated lung-sickness in the Okavango Native Reserve, and he was determined to replicate his success in Ovamboland. By the 1950s, however, increasingly frustrated in his efforts, Eedes had recast indigenous Ovambo cattle from a subsistence asset into

---

35 Kreike, "Architects of Nature", pp. 111–113.

36 NAN, NAO 101, Senior Agricultural Officer Natives, Agricultural Survey Ovamboland, Windhoek, 26 Oct. 1947.

37 NAN, NAO 60, Quarterly Report Ovamboland, July-Sep. 1949.

a dangerous and contagious liability. After alleged outbreaks of foot and mouth disease in the late 1940s, his administration created the Red Line, a veterinary cordon between Ovamboland and the remainder of Namibia to its south.[38]

Eedes' measures intersected with 1940s and 1950s Portuguese attempts to create an indigenous cattle-breeding industry in south-central Angola, which led to large-scale cattle vaccination projects. Cattle epidemics in the 1940s and 1950s on both sides of the border consolidated an image of Ovambo cattle as a source of disease and an obstacle to development, resulting in further constraints on cattle movements across the Angolan-Namibian border.[39]

While most colonial experts and officials highlighted the 'explosive' increase in cattle numbers as Nature run amuck, other colonial sources paradoxically stressed that the high annual local livestock take-off due to consumption, disease and drought posed severe limits to growth.[40] Reports on beef consumption estimated that between 1966 and 1975 approximately 20–30 % of the total herd was slaughtered annually. For every five head of cattle that Ovambo households 'disposed of', roughly one head was consumed by the household and four were sold.[41]

Households that slaughtered cattle consumed part of the meat and sold or dried the remainder, and cattle and beef were widely traded, even across the Angolan border. In 1956, fourteen more or less formal 'native' butcheries operated in the Oukwanyama subdistrict of Ovamboland alone.[42] In 1942 (a year of severe drought), a live ox cost eight pounds sterling (sixteen rand) while in 1954, official prices in the Union of South Africa for a good quality head of Ovambo cattle varied from two to seven pounds (four to fourteen rand). Cattle and beef readily

---

[38] NAN, NAO 15, Veterinary Officer to Director Agriculture, Investigation: Foot and Mouth Disease, Jan. 18, 1946; NAO 106, Diary NCO 1949–1954, entry 13 Feb. 1950; OVA 49, Chief Agricultural Officer to Director Agriculture, Ondangwa, 25 June 1969.

[39] Kreike, *Re-creating Eden*, pp. 171–176. See also chapter 3.

[40] NAN, AHE (BAC) 1 / 352, Annual Reports Agriculture Oukwanyama 1964 and 1968; OVA 49, Meeting of the Sub-Committee on Village Planning, 2 Sep. 1970; OVA 9, Statistics 1967, appendix to Director-in-Chief Economic Affairs to Director Agriculture, Ondangwa, 25 March 1969, and Director Agriculture to Director-General Cooperation Pretoria, [Ondangwa], 5 May 1981; OVA 61, Monthly Reports Agricultural Officer: Andreus Ndeitwa, Aug. 1976; OVA 6, Annual Report Agriculture Owambo, 1979–1980. On the 'backwardness' of Ovambo cattle management, see AHE (BAC) 1 / 346, Chief Bantu Commissioner SWA to Principal Agricultural College Arabie, Transvaal, Windhoek, 13 July 1965.

[41] Kreike, "Architects of Nature", p. 114.

[42] NAN, BAC 133, Agricultural Report Ovamboland, 1955–1956.

were exchanged for the staples millet and sorghum.[43] Cattle hides were also sold. During drought years in the 1940s and 1950s, cattle owners from Ovamboland annually smuggled 5–10,000 tanned hides to Portuguese traders across the border, receiving one pound sterling apiece. Although formally prohibited, the export of cattle hides to southern Angola continued until at least the early 1970s.[44]

The sale of cattle and cattle products was not the only factor that limited cattle population growth. Estimated losses due to drought and disease were 30 % in 1941, 5 % (12,500 head) in 1945–1946, and 10 % (40,000 head) in 1979–1980.[45] With annual consumption levels as high as 20–30 %, combined with years with disease- and drought-induced losses of up to 5–30 %, it seems remarkable that there would be any livestock left at all, let alone that there would be any increase. As is the case with the statistics for the actual number of cattle, however, these figures are estimates. More important is that the consumption, sale and loss numbers overlapped because the owners consumed the meat of perished livestock and sold the hides. In addition, cattle owners routinely slaughtered drought- or disease-weakened animals.[46]

Still, the figures demonstrate that at the very least, some colonial officials had the impression that cattle and cattle products were important consumption and trade goods within Ovamboland and across the Angolan border. In other words,

[43] NAN, NAO 98, ANC to NCO, Oshikango, 11 June 1947, and statement Mululu Kalongela, Ondangwa, 12 May 1948; NAO 64, Minutes of Ukwanyama Tribal Meeting [12 July 1954]; AGR 897, statement Elizabeth Ikau, 27 Nov. 1961; BAC 133, Agricultural Report Ovamboland, 1955–1956. For other examples of the exchangeability of cattle and grains, see NAO 98, ANC to NCO, Oshikango, 11 June 1947, and statement Mululu Kalongela, Ondangwa, 12 May 1948; A450, 23 D4 (1924) and 24 D19.

[44] NAN, BAC 133, Agricultural Reports Ovamboland, 1955–1956 and 1956–1957; NAO 27, NCO to Sec. SWA, Ondangwa, 15 Nov. 1941; NAO 103, Censuses of Agriculture Ovamboland, 1945–1946 and 1949–1950; NAO 70, statement Nehala Nangoro, Oshikango, 23 Dec. 1948; NAO 64, Minutes of Ukwanyama Tribal Meeting [12 July 1954]; BAC 122, Famine Relief Schemes, Chief Bantu Commissioner to Minister for Bantu Administration, Windhoek, 26 June 1959; BAC 44, minutes meetings, Ombalantu, 15 June 1960 and Onkolonkathi, 16 June 1960; OVA 50, Minutes Ukuanyama Tribal Government Meeting, 7 Dec. 1971.

[45] NAN, BAC 133, Agricultural Reports Ovamboland, 1955–1956 and 1956–1957; NAO 103, Census of Agriculture Ovamboland, 1945–1946, and NCO to Sec. SWA, Windhoek, 8 March 1946; BOS f. "Oshikango", Agricultural Officer Ovamboland to Native Commissioners Ondangwa and Oshikango, [Ondangwa], 17 Aug. 1956; OVA 6, Annual Report Agriculture Owambo, 1979–1980; A450, 7, Annual Report Ovamboland 1941.

[46] On eating dead cattle, see NAN, NAO 60, Quarterly Reports Ovamboland, Jan. – March 1948, Jan. – June 1949, Jan. – March 1951, Jan. – March 1954; NAO 59, Kaibi Mundjele to NCO, Ombalantu, 2 May 1950; NAO 37, Annual Health Report Ovamboland 1937; NAO 20, Monthly Reports Ovamboland, March – April 1940; NAO 62, Agricultural Report Ovamboland 1953.

cattle retained a substantial level of commodification locally, despite having been decommodified at the level of the formal colonial and international markets after the imposition of colonial rule in 1915.[47]

Colonial barriers: Conservation and fences

In the 1950s, 1960s and 1970s, vaccinations and fences became the primary mechanisms for containing Ovamboland's presumed livestock problem. The objectives were to conserve Ovamboland's environment and to protect from contamination the livestock in the white settler farming areas to the south of Ovamboland, in Angola and in neighboring Kaokoland and Okavango, and to protect the wildlife in Etosha Park. Considerations of scientific management thus legitimized and institutionalized the idea of confining Ovamboland's cattle resources to an increasingly smaller space, inhibiting the cattle's mobility and access to dry season cattle posts and perpetuating the animals' decommodification in any markets outside of the reserve.

In the 1960s and 1970s, the colonial administration erected fences along the Angolan border, the western Kaokoland border and the Red Line to the south of Ovamboland in order to contain diseases and to prevent the re-infection of vaccinated animals by unvaccinated livestock and wild animals. Most of the Etosha Park boundary with Ovamboland proper was fenced between 1971 and 1974. Although the fences proved to be far from 'cattle-proof' or 'game-proof', they seriously hampered seasonal animal movements.[48] The impact of colonial fencing on livestock management in Ovamboland was dramatic because it cut herdsmen

[47] In western Kenya similar local cattle markets operated below the colonial radar screen, see Cohen and Atieno Odhiambo, *Siaya*, pp. 76–81.

[48] NAN, AGR 125, Director Agriculture to Director Veterinary Services Pretoria, [Ondangwa], 27 Aug. 1959; BOS, "District Record Book Oshikango", 1965; AGR 897, Director of Agriculture to Cattle Inspector Oshikango, [Ondangwa], 8 Nov. 1961, and to Secretary SWA, [Ondangwa], 13 Jan. 1961; BAC 40, Director Agriculture to Chief Bantu Affairs Commissioner, Windhoek, 15 March 1963; AGR 95, Veterinary Inspector to Director Agriculture Windhoek, Omafo, 15 May 1963; AHE (BAC) 332, State Veterinarian to Director Agriculture, Ondangwa, 23 Sep. 1966; AGR 298, memo Director Veterinary Services to Secretary OTC, 19 May 1968 and Director Agriculture to Administrator SWA, 1 Nov. 1968; OVA 56, Chief Bantu Commissioner SWA to Chief Director Ovamboland, 18 Dec. 1968, appendices I–II.

and herds off from some of the most important dry season forages beyond the Ovambo floodplain. The fences changed the transhumance system and the population's access to cattle and cattle products.

Transhumance treks became longer during the 1930s, 1940s and 1950s, even before the fences were erected, for at least two reasons. First, Portuguese officials often interfered with herders who engaged in cross-border transhumance to the cattle posts along the Kunene and Kavango rivers and in Oshimolo. After the delimitation of the Angolan-Namibian colonial boundary, however, herdsmen from the villages south of the border continued to take their animals across the border to cattle posts along the Kunene River and to Oshimolo. Second, the expanding village landscapes of Ovamboland encroached on the uninhabited wilderness of the middle floodplain and beyond, including in areas that formerly had been sites for cattle posts. As a consequence, cattle were kept away from the villages for increasing periods, and critical cattle products (manure, dairy products and meat) were less readily available. By the mid-1950s and 1960s, a household's cattle often remained at remote cattle posts during the entire dry season.[49] In August 1966, livestock kept near villages reportedly was losing "condition" while livestock in the less densely inhabited areas remained in good "condition".[50] Herding the cattle back to the villages after the rainy season began was a top priority even in the face of veterinary restrictions. In early 1967, for example, temporary veterinary cordons along the Angolan-Namibian border to contain an outbreak of cattle diseases proved impossible to maintain because owners persevered in bringing their cattle back from the Angolan cattle posts to their Namibian homesteads.[51]

Before the 1960s, Ovamboland's cattle itself were identified as a severe obstacle to pastoral development because they were considered to be a poor, disease-ridden species. But new research in the late 1960s and 1970s demonstrated that the region's indigenous cattle were so well adapted to the extreme environmental

---

[49] NAN, BAC 133, Agricultural Report Ovamboland 1955–1956; BAC 132, Trust Farming Projects, Agriculture Officer to Bantu Commissioner, Grootfontein, 1 Feb. 1962, Monthly Report for Jan. 1962; Kreike, *Re-creating Eden*, chap. 8.

[50] NAN, AHE 1/351, Report of Activities Agriculture Ovamboland, July-Sep. 1966 and monthly report Aug. 1966.

[51] NAN, AGR 47, Director Agriculture to Director Animal Research Institute Pirbright, England, n.p., 23 Feb. 1967.

conditions there that they were potentially more efficient reproducers and beef producers than any other breed. In addition, the research suggested that selective breeding had the potential to further improve the breed.[52] When the administration's measures to enhance veterinary services and propagate modern cattle management techniques faltered in the 1970s and 1980s, however, colonial officials and experts even more explicitly than previously blamed the failure on the 'primitive' nature of Ovamboland's cattle holders, claiming that the latter simply refused to cooperate with the administration's selective breeding and grazing rotation programs, or to offer their animals for sale at the new Development Corporation's meat-processing plant at Oshakati. Indeed, between its opening in 1976 and 1981, the Oshakati plant purchased over 50,000 head of cattle, but only 500 came from Ovamboland's cattle holders, who were the intended suppliers. White ranchers supplied the bulk.[53]

Given the circumstances, it is somewhat unsurprising that a 1986 report phrased the challenges to the development of colonial Namibia's livestock sector in terms of a cattle complex: "In traditional areas cattle are considered to be a cultural asset of the community and as such it is difficult or even impossible to exploit cattle farming commercially".[54] Keith Morrow, the author of the 1988 report that opened this chapter, however, blamed the policies of colonial Namibia's Department of Agriculture as well as the 'primitive' attitudes of Ovamboland's inhabitants. For example, on the one hand, Morrow chided the parastatal Development Corporation for offering below-market prices to its potential

[52] NAN, AGR 538, Director Agriculture to Bishop of Damaraland, Windhoek, 26 May 1961; OVA 40, Ovambo Government, Verhoging van die Beesvleis produksie in Ovambo deur seleksie en teling, 5 Aug. 1971; OVA 26, H. Hamburger, C.W.B. Armstrong and J. Swanepoel, "Adaptability and Reproductive Efficiency: The Value of Indigenous Sanga Cattle in the National States of South Africa and Namibia", Republic of South Africa: Department of Co-operation and Development [1979 or 1980?].

[53] NAN, AHE (BAC) 1/352, Annual Report Agriculture Oukwanyama for 1964 and Annual Report for Agriculture Ovamboland for 1968; AHE (BAC) 1/346, Chief Bantu Commissioner SWA to Principal Agricultural College Arabie, Marble Hall, Transvaal, Windhoek, 13 July 1965, and Travel Report Chief Agricultural Officer, April 25–27 1966; AHE (BAC) 332, State Veterinarian to Director Agriculture, Ondangwa, 23 Sep. 1966; OVA 49, Meeting of the Sub-Committee on Village Planning and Development and Agricultural Planning of the Planning and Coordinating Committee on 2 Sep. 1970; OVA 40, Gresse to Director Agriculture, Ondangwa, 30 March 1971 and travel report 8–12 March 1971; OVA 9, Director Agriculture to Director-General Cooperation and Development Pretoria, [Ondangwa], 5 May 1981.

[54] Proposed Agricultural Strategy for SWA / Namibia, Aug. 1986.

Ovambo sellers; on the other hand, he attributed overgrazing in Ovamboland to the drive to maximize cattle numbers to enhance individuals' social status.[55]

Grazing pressure and desertification

The concept of carrying capacity is based on the presumption that capacity to sustain ('carry') livestock of any given amount of land has natural limits, an idea that is expressed in terms of how many hectares are required to sustain a single head of livestock. Colonial research during the 1970s concluded that western Ovamboland could support 1 Cattle Unit (CU), i.e., one head of cattle, per 12 ha. Central and eastern Ovamboland had a carrying capacity of only one CU per 16 ha. Actual grazing pressure in central Ovamboland, however, was thought to be as high as one CU per 3.5 to 5 ha, four times the carrying capacity. In 1974, the carrying capacity of Ovamboland as a whole was set at 463,941 CU, but the total livestock population was 564,135 CU. In 1977, based on a 1975 count, Ovamboland's 5,567,400 ha were thought to support 660,704 CU, giving a stocking rate of one CU per 8.14 ha instead of the optimal one CU per 12–16 ha derived from the research mentioned above.[56] As a result, desertification was thought to be imminent. According to one report, "The soil destruction (…) in the densely populated parts of Ovamboland is shocking and if this degradation is not halted soon, it can not be prevented that large parts of the area [Ovamboland] will be transformed in a unhospitable desert".[57] Consequently, during the early 1970s, colonial extension workers strongly urged livestock owners to take action against overstocking and overgrazing by adopting modern management practices, including grazing rotations and selective cattle breeding.[58]

---

55 [Keith Morrow], A Framework for the Long Term Development of Agriculture within Owambo (Aug. 1989).

56 NAN, OVA 26, Sec. Agriculture to Sec. Bantu Administration Pretoria, Ondangwa, 10 March 1977.

57 NAN, BAC 133, Agricultural Report Ovamboland 1955–1956; OVA 57, Dr. H.A. Lueckhoff, report on a visit to South West Africa, 3–15 Nov. 1969, appendix to Regional Forester to Director-in-Chief Bantu Administration Pretoria, Grootfontein, 3 April 1970; OVA 56, Sec. Bantu Administration to Director Agriculture Ondangwa, Pretoria, 11 June 1971; OVA 46, Director Agriculture to Sec. Bantu Affairs Pretoria, Ondangwa, 17 Nov. 1971; OVA 45, Sec. Agriculture to Foreign Affairs Pretoria, Ondangwa, 16 Jan. 1974.

58 See, for example, NAN, OVA 61, Agricultural Officer Moses Nadjebo, Monthly Reports 1971.

However, the assessment of severe overstocking and the conclusions derived from that analysis are problematic at best. Estimates of the carrying capacity of Ovamboland varied considerably. For example, although in the early 1970s the acceptable carrying capacity for Ovamboland was set at one CU per 12–16 ha, a mid-1980s assessment set it much lower at one CU per 8–10 ha, allowing for a much higher livestock population.[59] The very concept of carrying capacity is in itself controversial.[60] Moreover, presumptions about the grazing impact for South Africa and colonial Namibia were based on trials with Afrikaner and exotic cattle. The indigenous Sanga cattle, despite their lower body mass, were not differentiated from such heavier cattle breeds as Afrikaner. Yet, an overall 10 to 20% lower body mass for Sanga cattle should result in a correspondingly lower environmental impact. Of the 560,000 CU in the previously mentioned report, 488,000 consisted of cattle and 72,000 of other stock.[61] Based on a Sanga-adjusted 0.8 CU, for example, the CU aggregate comes to 468,000, which is very close to the 1974 figure for Ovamboland's carrying capacity of 463,941 CU. Thus the colonial carrying capacity data offer little or nothing to support the assertion that overgrazing was a major cause of environmental degradation before the mid-1970s. Moreover, during the 1980s, the cattle population of Ovamboland declined.

Livestock and deforestation

In the 1970s and 1980s, foresters and other experts perceived the relationship between livestock and 'forest' in Africa in negative linear terms: increased

---

59 See Proposed Agricultural Strategy for SWA / Namibia, Aug. 1986.

60 Sanford, *Management of Pastoral Development in the Third World*, pp. 104–105; Little, "Rethinking Interdisciplinary Paradigms and the Political Ecology of Pastoralism in East Africa", and Munro, "Ecological 'Crisis' and Resource Management Policy in Zimbabwe's Communal Lands".

61 NAN, OVA 45, Sec. Agriculture to Foreign Affairs Pretoria, Ondangwa, 16 Jan. 1974; OVA 26, H. Hamburger, C.W.B. Amstrong and J. Swanepoel, "Adaptability and Reproductive Efficiency: The Value of Indigenous Sanga Cattle in the National States of South Africa and Namibia", RSA, Department of Co-operation [c. 1979]; OVA 55, le Roux, "A Progress Report on Indigenous Cattle in SWA", 20 June [c. 1980]; BOS f. "Oshikango", Agricultural Officer Ovamboland to Native Commissioners Ondangwa and Oshikango, [Ondangwa], 17 Aug. 1956. A 1972 report specified that the CU standard was based on a head of cattle of 800 lb., see AGR 541, Director [Nature Conservation?] SWA to Sec. Agriculture, Technical Service Pretoria, n.p., 16 Oct. 1972.

livestock numbers caused increased deforestation.[62] In Ovamboland as in other semi-arid and arid environments across the globe, woody vegetation was and continues to be an important dry season source of browse for cattle and other livestock.[63] Towards the end of the dry season, as the vegetation dries, the grasses' nutritional quality rapidly drops. In Southern African parlance, the grasses turn 'sour'. Browse from trees and bush, including mopane leaves, remains easier to digest and loses mineral content much more slowly.[64] Moreover, trees and bushes turn green before grasses and herbs at the turn from the dry to the rainy season.[65]

Foresters feared that livestock browsing on trees in north-central Namibia would result in deforestation and soil degradation.[66] Yet, evidence of livestock-induced environmental degradation is hard to pinpoint. First, north-central Namibia today is not a treeless desert; far from it. Moreover, two indicators of degradation caused by overgrazing that are widely used are rarely employed in the case of Ovamboland. For example, while bush encroachment has not been identified as an issue in Ovamboland, it is cited as a major problem in the white settler ranching areas to the south as well as in the former Native Reserves in Namibia and South Africa. In addition, veterinary officials identified poisonous plants as being a problem in Ovamboland. The mere presence of such plants, however, is an unreliable proxy for overgrazing because even if the plants truly were nefarious invaders, herdsmen can prevent livestock from eating them and the animals apparently learn to avoid them. An increase in the incidence of plant poisoning is thus a poor indicator of vegetation degradation because it may result

---

[62] Le Houérou, *The Grazing Land Ecosystems of the African Sahel*, pp. 90–128; H. Gillet, "Observations on the Causes of Devastation of Ligenous Plants in the Sahel and Their Resistance to Destruction", in Le Houérou, *Browse in Africa*, pp. 127–129; Westoby, *Introduction to World Forestry*, pp. 172–173.

[63] See Le Houérou, *Browse in Africa*, p. 3, and the contributions there by Walker, "A Review of Browse and Its Role in Livestock Production in Southern Africa", p. 12; McKell, "Multiple Use of Fodder Trees and Shrubs"; and Le Houérou, "The Role of Browse in the Management of Natural Grazing Lands".

[64] This aspect is discussed in more detail in Kreike, "De-Globalization and Deforestation in Colonial Africa".

[65] Personal observations, 1991–1993, and Walker, "A Review of Browse and Its Role in Livestock Production in Southern Africa", p. 16, table 9. For West Africa, see Le Houérou, *The Grazing Land Ecosystems of the African Sahel*, pp. 49–50, 82.

[66] NAN, OVA 57, Le Roux to Sec. Agriculture, Supply Inventory: Indigenous Forests Ovamboland, Ondangwa, 5 Nov. 1976.

from a decline in the quality of herd management as opposed to stemming from overgrazing per se. This was clearly the case in Ovamboland as the demand for migrant labor shifted the burden of herding cattle in the villages and cattle posts to increasingly younger and less-experienced boys.[67]

Commodification, deglobalization and deforestation

South African colonial officials' and experts' concern that the explosive growth of the cattle population in Ovamboland would set the territory on a destructive course towards overgrazing, deforestation and desertification predates World War II. Yet, despite a surge in cattle population beginning in the early 1920s, by 1990, north-central Namibia was neither a treeless desert, nor bush-encroached, nor was it overrun by poisonous plants. Moreover, even if colonial cattle statistics are compared with the colonial concept of carrying capacity to quantify the environmental impact of Ovamboland's cattle, the outcome is ambiguous, and it certainly does not support the conclusion of dramatic overstocking.

Paradigms of environmental change that are derived from a Nature-Culture dichotomy typically use as their point of departure a local, precolonial, state-of-Nature subsistence economy that is threatened by an invasive colonial global market economy. But the 1890–1990 history of Ovamboland inverts (and subverts) the Nature-to-Culture narrative. The region's cattle were a global commodity before colonial conquest in 1915, and colonial rule transformed the region's cattle from a global commodity into a local subsistence resource, although colonialism typically is regarded as an agent of market capitalism. Moreover, no global market-driven overexploitation of the region's indigenous cattle resources followed in the wake of colonial occupation as might have been theorized by a Marxist-inspired declinist model. While the colonial-era recovery of the cattle herds raised fears about a neo-Malthusian cattle population and disease time bomb, the specter of an imminent environmental catastrophe was exaggerated.

[67] Kreike, *Re-creating Eden*, pp. 155–176; NAN, NAO 59, Dr. Zschokke, Survey of Cattle Diseases in Ovamboland: Oct. 1948, 5 Nov. 1948, and OVA 6, Annual Report Veterinary Service Owambo 1975–1976.

Ultimately, colonial officials and experts did not attribute the rapid increase in cattle numbers to the prohibition against exporting cattle or to the introduction of modern health and management practices, which purportedly reduced mortality and caused the neo-Malthusian trap. Instead, they singled out the persistence of 'primitive' (precolonial) cattle management practices in Ovamboland: 'primitive' pastoralism was identified as the reason that cattle holders resisted, for example, veterinary programs to eradicate cattle diseases. By extension, 'primitive' pastoralism consequently forced the colonial administration to impose the Red Line and to prohibit the export of cattle from the reserve. Except for a short period in the 1980s, from 1890 to Namibian independence in 1990 South African colonial officials and experts on the ground did not invoke the classic cattle complex explanation that local cattle owners were resistant to marketing cattle. Rather, they formulated the more general argument that Ovamboland's cattle holders resisted modern scientific improvements because they were unwilling or unable to embrace the colonial technology that would help them to overcome the local environmental limits that were imposed by Nature.

Was the Ovamboland case unique? It seems unlikely. Cattle were a major export across the subcontinent before the consolidation of colonial rule. Lungsickness, rinderpest and war decimated cattle herds, causing crisis and famine throughout late nineteenth- and early twentieth-century Africa. Young men across the continent invested migrant labor wages in rebuilding the herds, even as colonial rule closed export markets. In the name of conservation and development, colonial governments in southern Africa imposed veterinary and other measures that restricted the movement of cattle and the ways in which people could manage and use them. Except in Botswana, indigenous cattle breeders remain cut off from global markets. Colonialism, therefore, neither unambiguously promoted a market economy transition from Nature to Culture, nor was colonialism an unequivocal harbinger of economic globalization and its attendant environmental consequences.

7

The Palenque paradox: Beyond Nature-to-Culture

The urban environment is often seen as the consummate cultural landscape, and a polar opposite of pristine Nature or wilderness. Aboriginal people such as, for example, the 'bushmen' or San of southern Africa frequently are viewed as indicators of unspoiled wilderness and their lifestyles are considered to be the antithesis of Western urbanism.[1] But the wilderness expanses of Central America, Africa and Asia contain the ruins of a large number of 'lost' cities.[2] The San and their environments similarly may have not been as 'wild' as sometimes has been assumed. At the same time that European travelers uncovered Palenque and Great Zimbabwe, the San of southwestern Africa engaged in activities that seemed in sharp contradiction to the behavior expected of archetypical nomadic Stone Age hunters and gatherers. Armed with the latest firearms technology,

[1] A good example of the popular image of simple 'Stone Age' life of 'Bushmen' and its juxtaposition with modern urban life is the film *The Gods Must Be Crazy*. Sahlins stresses that San / Bushmen and other similar communities rejected the "Neolithic revolution", in *Stone Age Economics*, p. 27. See also Lee, "What Hunters Do for a Living".

[2] For the 'discovery' of stone structures in Central America and Africa in particular, see, for example, Stuart and Stuart, *Lost Kingdoms of the Maya*, and Venning, "Notes on Southern Rhodesian Ruins in Victoria District". On Asia, see below, note 4.

they engaged in commercial hunting; they were involved in mining copper and iron; and they lived in permanent villages. Within the reigning paradigms of environmental change premised on a Nature-Culture dichotomy, the presence of urban ruins in the jungles and plains of Central America and Africa, and the degree to which San society was in fact embedded in Culture rather than Nature, are problematic. At the very least, these phenomena constitute a paradox.[3]

Bush cities and the bush

The 'jungles' or 'bush' of Central America, Southeast Asia and Africa are not only treasure troves of biodiversity, but also are home to some of the most impressive remnants of urban civilizations, including Palenque in Mexico, Angkor Wat in Cambodia and Great Zimbabwe in Zimbabwe.[4] Such national parks and reserves as South Africa's Kruger Park serve as the last sanctuaries of an African Eden. But conservation areas seldom contain pristine Nature. Many if not all of Africa's national parks and reserves were converted into 'wilderness' by forcibly removing the local populations and prohibiting them access to the local resources.[5] This phenomenon was not confined to Africa: clearing out populations and denying them access to forest reserves and other conservation areas in Asia especially has been marked by fierce resistance, frequently making conservation at best precarious.[6]

---

[3] For the classic study on the history of the concept of wilderness, see Nash, *Wilderness in the American Mind*. Cronon rejects what he calls the human-natural and urban-rural dichotomies and White points to the 'blurring' of the wild-domesticated dichotomy. See Cronon, *Nature's Metropolis*, pp. 17–19; and White, *The Organic Machine*, pp. 105–109. Merchant criticizes both the progressive (here: modernization) and declinist paradigms as linear and unidirectional, in *Reinventing Eden*, pp. 4, 6, 215. See also Pimentel, Westra and Noss, *Ecological Integrity*, pp. 7–8; and Huxley, *Tropical Agroforestry*, p. 301.

[4] On Great Zimbabwe, see Hall, *The Changing Past*, pp. 91–116; and Beach, *The Shona and Zimbabwe*, pp. 1–52. On Palenque, Stuart and Stuart, *Lost Kingdoms of the Maya*, pp. 19, 31, and Perera and Bruce, *The Last Lords of Palenque*, pp. 10–26. On Angkor Wat, see Audric, *Angkor Wat and the Khmer Empire*, and Chandler, *A History of Cambodia*.

[5] See Carruthers, *The Kruger Park*; Stevinson-Hamilton, *South African Eden*; and de Villiers, *Land Claims and National*. For East Africa, see Neumann, *Imposing Wilderness*.

[6] Guha, *The Unquiet Woods*, and Peluso, *Rich Forests, Poor People*.

Evidence of the cultural landscapes in what at present is defended as the last relics of pristine Nature is paradoxical: how can the site of the historical city of Palenque be pristine Nature? Two conceptions have somewhat eased the tension. One is the idea that premodern societies somehow 'lived by Nature', that they were sustained within the bounds and limits of their natural environment. This idea is expressed most explicitly in the literature about Native Americans, the San, the Batwa ('pygmies') and other 'aboriginal' hunters and gatherers.[7] On the other hand, however, indigenous settled agricultural societies are sometimes perceived as being environmentally destructive: theories about the decline of Central American urban societies, for example, attribute their demise to environmental overexploitation.[8]

The second conception is much more implicit. Urban vestiges in 'Wild' Africa and the 'rainforests' of Central America are depicted as though they are entirely isolated from the surrounding natural ecosystems. The most dramatic image is the view that greets visitors to Palenque: the white stone buildings contrast sharply with the towering green jungle that envelops them. The ruins seem an anomaly, vulnerable even, their frailty punctuated by machete-wielding crews of men who engage in an endless battle to keep the wilderness at bay. The human corruption of Nature thus seems quarantined. In addition, an archaeological bias towards the spectacular and the monumental led to a disregard for the mundane, including details of the daily functioning of residential sites and due appreciation for the size of the population.[9] Another bias that prevented recognizing (sub)tropical large-scale societies as 'urban' stemmed from their striking climatic, geographical and demographic traits as compared to societies in the temperate zones. Precolonial

[7] For critiques of the concept of a premodern human-nature "balance" in North America, see, for example, Isenberg, *The Destruction of the Bison*, esp. pp. 2–10. For a similar critique regarding Central America, see, for example, Wingard, "Interactions between Demographic Processes and Soil Resources,"; and MacLeod, "Exploitation of Natural Resources in Colonial Central America". Balée in the latter volume offers a countercritique and argues that Indians did not cause any dramatic biodiversity losses, "Indigenous History and Amazonian Biodiversity".

[8] Stuart and Stuart, *Lost Kingdoms of the Maya*, pp. 63–64, and Vlcek, Garza de Gonzalez and Kurjack, "Contemporary Farming and Ancient Maya Settlement: Some Disconcerting Evidence", esp. pp. 218–220.

[9] Hartland was one of the first to claim that the temples were the centers of cities and towns, in "Maya Settlement Patterns", p. 37. See also N.P. Dunning, "A Reexamination of Regional Variability in the Pre-Hispanic Agricultural Landscape", in Fedick, *The Managed Mosaic*, pp. 53–91. On the archaeological bias, see Graham and Prendergast, "Maya Urbanism and Ecological Change"; and Stuart and Stuart, *Lost Kingdoms of the Maya*, p. 32.

urban centers in the (sub)tropics often were subject to a climate with a sharp dry-rainy season variability. They also tended not to be located along major (permanent) rivers, and settlement may have been less concentrated than in temperate zones.[10]

In *Nature's Metropolis*, Cronon argues against the separation of urban and nonurban and of human and natural worlds. He demonstrates that urban Chicago and its rural and wilderness hinterland were part of a continuum, and that they were mutually dependent upon one another. Nineteenth-century Chicago's urban growth led to a cutover of the forest regions to the north and transformed the bison prairies to the west into wheatfields and cattle range.[11] In perhaps less spectacular ways, Central American, Southeast Asian and African urban centers similarly were connected to and dependent on large hinterlands.[12] Precolonial urban centers in the Americas were large and numerous. The Aztec capital of Tenochtitlan boasted 250,000 inhabitants, or five times the population of contemporary London when Cortés first laid eyes on it. Tikal in the eighth century was surrounded by expanses of suburbs. Nohmul's remains cover an area of 35 square kilometers. Precious little vacant land remained in the area immediately around the central pyramid group at Chunchumil; the site contained 2,400 housing complexes with a total estimated population of 12,000 on a mere 6 square kilometers. Mexico alone has 10,000 known sites of Meso-American cities and towns and Guatemala has another 1,400. Great Zimbabwe had a population of up to 30,000 and over 50 other associated sites that also have significant ruins.[13]

---

10 Scarborough, "Reservoirs and Watersheds in the Central Maya Lowlands". This idea is also very valuable in looking at African precolonial urbanization: Great Zimbabwe, for example, was also not located near a major river.

11 Cronon, *Nature's Metropolis*, pp. 17–19.

12 Graham and Prendergast argue that "pre-industrial tropical urbanism" sheds new light at the ways environmental change is studied, in "Maya Urbanism and Ecological Change". This hints at the issues at the heart of Cronon's argument, since the nature of precolonial 'urbanism' in Central America and Africa in many ways challenges the Eurocentric models of 'urbanism': for example, African and Central American urbanism may have been less nucleated and less discrete.

13 On Tenochtitlan's population and the number of Meso-American ruins, see Foster, *A Brief History of Central*, pp. 20–21. On Chunchumil, see Vlcek, Garza de Gonzalez and Kurjack, "Contemporary Farming and Ancient Maya Settlement", pp. 212–217. On Nohmul, see Pyburn, "The Political Economy of Ancient Maya Land Use", pp. 242–243. On Tikal, see Stuart and Stuart, *Lost Kingdoms of the Maya*, p. 32. On Great Zimbabwe, see Hall, *The Changing Past*, pp. 91–116.

Construction and maintenance of the urban structures and their populations must have been a severe drain on the hinterlands. The Incas had a centralized system that supplied urban centers with firewood originating from the Andean forests.[14] The stone used to construct Angkor Wat came from quarries 30 kilometers away. Maintenance was costly: 80,000 people in 3,140 villages were assigned to the upkeep of a single temple in Angkor Wat that was built by King Jayavarman VII in the twelfth century.[15] In addition, intricate networks of exchange connected urban centers and their hinterlands to one another. This was as true for Tikal, Palenque and Great Zimbabwe as for nineteenth- and twentieth-century Chicago. Great Zimbabwe traded as far southwest as Toranju, near the Makgadigadi pans in central Botswana and as far north as Ingombe Ilede in the Zambezi Valley, and it was a core area in an Indian Ocean trade system that stretched as far east as China.[16]

Food production requirements to sustain the urban populations also support the concept of an urban-rural continuum. Mayan cities and towns contained garden plots for intensive horticulture.[17] Much of the food, however, must have originated from rural areas. Before the late 1970s, pre-Columbian agriculture was thought to have been dominated by *milpa*, or temporary fields in a slash-and-burn cultivation system. Research in the 1970s and 1980s, however, found evidence of much more intensive forms of cultivation, based on raised fields and intricate systems of irrigation and water harvesting and storage, suggesting levels of environmental manipulation that contradict notions that pre-Columbian Indian populations simply lived by Nature.[18] In addition, fields were not necessarily close to

---

[14] Gade, *Nature and Culture in the Andes*, p. 52.

[15] Higham, *The Archaeology of Mainland Southeast Asia*, pp. 333, 339–340.

[16] On trade, see Vlcek, Garza de Gonzàlez and Kurjack, "Contemporary Farming and Ancient Maya Settlement", pp. 222–223; and Puleston, "Terracing, Raised Fields, and Tree Cropping in the Maya Lowlands", p. 244; Dunham, "Resource Exploitation and Exchange among the Classic Maya"; Foster, *A Brief History of Central America*, pp. 17, 28, 36–37. On Great Zimbabwe, see Hall, *The Changing Past*, pp. 91–116.

[17] For in-town garden plots, see Vlcek, Garza de Gonzalez and Kurjack, "Contemporary Farming and Ancient Maya Settlement", pp. 212–217; and Puleston, "Terracing, Raised Fields, and Tree Cropping", pp. 226–227, 229–230. See also Dunning, "A Reexamination of Regional Variability in the Pre-Hispanic Agricultural Landscape". Dunning defines the ancient communities in the Puuc hill country of Yucatán and Campeche as "garden cities".

[18] Graham and Prendergast, "Maya Urbanism and Ecological Change", pp. 102–109; and Hoffmann, "La importancia ecológica y económica de las tecnologías tradicionales en la agri- y silvicultura en áreas de bosque tropical humido en México"; Puleston, "Terracing, Raised Fields, and Tree

the main residences.[19] Minerals and forest products were acquired even beyond the agricultural hinterlands of the cities and towns. Upland forests, for example, were long considered to be of minor importance to the Maya, but they were relatively close to urban centers. The upland forests 100–150 kilometers from Tikal, in fact, were the source of many sought-after minerals, and wherever water was available, sizable settlements were established to extract forest resources.[20] The evidence suggests that much of Meso-America may have been a domesticated environment. Indeed, the seemingly 'pristine forests' of the Maya lowlands are less than four hundred years old.[21] 'Primordial' forest wilderness in Latin America such as that found at La Selva research station in Costa Rica revealed human use as evidenced by pre-Columbian sites with charcoal and ceramic remnants.[22] In brief, every forested landscape bears evidence of human interventions over the course of the millennia.[23]

The above is even more true for Africa, which is considered the 'wild' continent par excellence. Africa's game parks and nature preserves are rarely pristine; most of the protected areas have long histories of human use. As anyone who has flown between Johannesburg and Maputo can attest, the southern Kruger Park is pockmarked by the remnants of human settlements. When much of the

---

Cropping in the Maya Lowlands"; and Matheny, "Northern Maya Lowland Water Control Systems", pp. 195, 205. For water harvesting, see also Scarborough, "Reservoirs and Watersheds in the Central Maya Lowlands". For raised fields and irrigation works in the pre-Columbian Andes highlands and lowlands, see Denevan, *Cultivated Landscapes of Native Amazonia and the Andes*. For "the myth of the *milpa*", see Fedick, *The Managed Mosaic*, p. 2. For the degree of environmental management involved in the case of Angkor Wat, see Frédéric, *La vie quotidienne dans la Péninsule indochinoise à l'Epoque d'Angkor*, 800–1300, pp. 137–138; Higham, *The Archaeology of Mainland Southeast Asia*, pp. 321, 325, 329, 337, 341, 348–352; and Audric, *Angkor and the Khmer Empire*, pp. 125–132.

[19] Vlcek, Garza de Gonzàlez and Kurjack, "Contemporary Farming and Ancient Maya Settlement", p. 220.

[20] Dunham, "Resource Exploitation and Exchange among the Classic Maya", pp. 320–325.

[21] Leyden, Brenner, Whitmore, Curtis, Piperno and Dahlin, "A Record of Long- and Short-Term Variation from Northwest Yucatán: Cenote San José Culchacá". Fedick stresses that ancient Maya agriculture supported millions of people over centuries in a marginal environment, in *The Managed Mosaic*, p. 10. See also Wingard, "Interactions between Demographic Processes and Soil Resources in the Copàn Valley, Honduras".

[22] Pierce, "Environmental History of La Selva Biological Station", esp. pp. 47–48. See also Konrad in the same volume, who argued that the dense forests of the Mayan peninsula formally came to be seen as "uninhabited" only since the late 1800s, "Tropical Forest Policy and Practice during the Mexican Porfiriato", p. 124.

[23] Rietbergen, *The Earthscan Reader on Tropical Forestry*, pp. 1–2; and Boyce, *Landscape Forestry*, p. vii.

South African Lowveld along the Mozambican border was transformed into what became the Kruger Park, some of the inhabitants were expelled while others initially were retained as gamekeepers and other staff. The Makuleke were forcibly expelled from the Parfuri Triangle in the far northern part of the park during the 1960s. Thus, although the first game warden of the Kruger Park propagandized the park as an African Eden, it was hardly pristine Nature.[24]

Wilderness in another remote corner of southern Africa similarly could not be considered to be 'pristine'. In the late 1800s, the Okalongo region in the middle Ovambo floodplain was described as an uninhabited hunters' paradise. Yet, when refugees from the northern Ovambo floodplain settled the area in the 1920s and 1930s, they encountered evidence of the presence of earlier settlement: fruit trees, water holes, water reservoirs and pottery fragments. They attributed these relics of a past humanized landscape to the prosperous kingdom of Haudanu, which was abandoned after a destructive war in the early 1800s. By the 1960s, the refugees and their descendents had once more transformed the Okalongo wilderness into a thriving and densely settled village landscape.[25]

Whereas war and population dislocation led to the recolonization of the Okalongo wilderness, the same processes had an opposite effect just north of Okalongo on the Angolan side of the border with Namibia. By the 1990s, local inhabitants designated large areas directly north of the Angolan-Namibian border on the western side of the Ovambo floodplain as wilderness.[26] Yet, less than a century earlier, the same areas had been described as lush, fruit tree-shaded expanses of adjoining farms and fields.[27] Indeed, the early twentieth-century northern Ovambo floodplain that supported the Ombadja and Oukwanyama kingdoms

---

[24] Adams and McShane, *The Myth of Wild Africa*, pp. 1–13; Neumann, *Imposing Wilderness*. On the Kruger Park, see Carruthers, *The Kruger Park*. On the Makuleke and the forced removals, see de Villiers, *Land Claims and National Parks*, pp. 45–57.

[25] See interviews by author: Julius Abraham, Olupito, 15 and 16 June 1993; Petrus Shanika, Oshiteyatemo, 17 June 1993; and Mathias Malaula, Onandjaba, 15 June 1993. See also E. Kreike, *Re-creating Eden*, chaps. 2, 4, and 7.

[26] See interviews by author: Julius Abraham, Olupito, 15–16 June 1993; Petrus Shanika, Oshiteyatemo, 17 June 1993; and Mathias Malaula, Onandjaba, 15 June 1993. Kreike, *Re-creating Eden*, chaps. 2, 4, and 7.

[27] For the Ombadjas, see especially Lima, *A Campanha dos Cuamatos*. See also Wülfhorst, *Von Hexen und Zaubern*, pp. 6, 13, and 17; AGCSSp, Duparquet 1879 journal, entries Aug. 12–14, 19–20, and 5 Sep. 1879; Veth, *Daniel Veth's Reizen in Angola*, pp. 340–341; Möller, *Journey in Africa*, pp. 110–112.

was amongst the most densely settled areas of what is modern Angola. The estimated population of pre-World War I Oukwanyama was from 60,000 to 120,000 inhabitants. The estimated population of the smaller Ombadja kingdoms, located on the western side of the northern floodplain, may have been between 15,000 and 60,000 inhabitants.[28]

Warfare, famine and disease associated with colonial conquest, and heavy taxation and forced labor between 1900 and 1930, led to massive mortality and flight, decimating the northern floodplain populations and triggering the abandonment of entire villages and districts, especially in the southern parts of the Ombadja region. By the 1930s, many of the formerly most densely populated Ombadja districts were uninhabited and entirely overgrown by bush vegetation. A Portuguese official report from 1919 claimed that the pre-1915 population density in Oukwanyama was 8 persons per square kilometer and for the Ombadjas, 12 persons per square kilometer. The figures may be inflated because they were used to legitimize indemnity claims against Germany, but they nevertheless illustrate that the region was considered to be fairly densely settled. In the context of the descriptions of the kingdoms as expanses of farms, fields and villages with an extensive water infrastructure, these reports emphasize the extent to which barely a century earlier, the 1990s wilderness had appeared to be very much like a semi-urban environment, or, at the very least, a 'managed mosaic' environment similar to that of 'rural' lowland Central America.[29]

'Bushmen' and the bush

Even the 'last wildernesses' that are home to 'Stone Age' hunter-gatherers, the San and Batwa, are not true wilderness.[30] San communities in the late nineteenth-

---

[28] AGCSSp, Duparquet 1879 journal, 8 Aug., 9 and 10 Sep. 1879; Duparquet 1882 journal, 14 July 1882; Ferreira Diniz, *Negocios Indigenas*, pp. 14–15. For a more detailed discussion, see Kreike, *Re-creating Eden*, chap. 3.

[29] Luiz de Mello e Athayde, "O Perigo de Despovoamento de Angola", pp. 229–230, AHU, Sala 8, Praca 115, Angola, Prejuizos causados pela guerra de 1914–18, Luanda, 14 Aug. 1919. For more details, see Kreike, *Re-creating Eden*, chaps. 3–4. Kjekhus describes similar processes in late nineteenth-century Tanzania, in *Ecology Control and Economic Development in East African History*, pp. 126–180.

[30] See, for example, Vansina, *Paths in the Rainforest*, p. 46.

and early twentieth-century border region in south-central Angola and north-central Namibia significantly shaped their environment. The San communities in nineteenth- and twentieth-century Botswana and Namibia were not isolated, small-scale, subsistence, nomadic hunter-gatherer societies. Namibian 'bushmen' controlled Tsumeb copper mining and engaged in commercial hunting.[31] San communities in the Mupa area were involved in iron smelting, and in central African myths the Batwa were associated with fire and iron, links that further undermine the image of the Batwa ('pygmies') and San ('bushmen') as Stone Age relics living by Nature.[32] Ovambo kings supplied ammunition and the most advanced firearms to their 'bushmen' business partners and the king and the hunter each received half of the ivory.[33] To qualify the region's San as Stone Age subsistence hunter-gatherers seems untenable given the significant evidence that they were entrepreneurs involved in mining and in commercial elephant hunting in addition to hiring themselves out to Ovambo kings as bodyguards and executioners.[34]

Moreover, the San who lived in or adjacent to the Ovambo floodplain may have contributed significantly to the demise of elephant herds in southern Angola and northern Namibia during the second half of the nineteenth century (and probably to the overhunting of other game) because they were the main elephant hunters in the region.[35] The decimation of the region's elephant population in the late nineteenth century, which was hastened by the 1897 rinderpest epizootic, coincided with a dramatic decline in the fortunes of the region's San communities – another indicator of how dependent the San had become on the commercial hunt and the global market in game products. By the early twentieth century, San inhabitants of the region sometimes quite literally had become the hunted instead of the hunters.[36]

---

[31] Wilmsen, *Land Filled with Flies*, and Gordon, *The Bushmen Myth*, pp. 15–43. For a classical study of the San, see Lee, *The !Kung*.

[32] Herbert, *Iron, Gender, and Power*, p. 63, and Klieman, "*The Pygmies Were Our Compass*", pp. 137–139.

[33] Kreike, *Re-creating Eden*, chap. 2.

[34] In 1928, King Martin of Ondonga in the southeastern floodplain employed San as hunters, messengers, and spies, NAN, NAO 18, Annual Report Ovamboland 1928.

[35] Cf. Isenberg who argues in *The Destruction of the Bison* that the Plains Indians contributed to the demise of the bison, and Foster who links the extinction of large mammals in Central America to Indian overhunting, *A Brief History of Central America*, p. 10.

[36] Kreike, *Re-creating Eden*, chaps. 2–3, 7.

As the northern floodplain war refugees transformed the 'wilderness' into landscapes of farms, fields and villages, colonial officials voiced concerns that the middle floodplain environment was becoming overpopulated and was threatened by deforestation and desertification. Unwilling to stop in-migration from Angola because of the increasing demand for migrant labor from Namibian and South African white farms, mines and industries, they cast their eyes on the area east of the Ovambo floodplain and south of the Angolan border: a land they deemed to be so inhospitable that it was said to be devoid even of flies. A 1928 report that highlighted the barrenness of the wilderness east of the Ovambo floodplain suggested that the San must have fled there: "he ['the' 'bushman'] has certainly not voluntarily chosen the Sandveld for his home".[37] The presence of the San was acknowledged only to confirm that the area was consummate 'wilderness', pristine, untamed and therefore potentially suitable to settle new refugees from Angola as well as migrants from the 'overpopulated' Ovambo floodplain.[38]

In the eyes of 1920s colonial officials, the San 'wildness' was underscored by acts of violence. This perception may in part have been a reflection of increased competition between different groups of San and between them and Ovambo hunters, herdsmen or migrants who moved into the wilderness east of the floodplain. In 1927, San reportedly attacked a group of thirty returning Angolan migrant laborers near the border.[39] In 1928, an expedition sent by the colonial administration to survey the east and to dig water holes was forced to turn back because, as its leader explained, "his natives were afraid of the Bushmen".[40] In 1929, a group of San was reported to have robbed two groups of migrant laborers from Angola, killing several. When King Martin of Ondonga sent an armed detail to investigate the incidents, his men clashed with the San perpetrators in their far eastern hideout, killing their leader. King Martin's men also reported "that the Bushmen had a fight amongst themselves over certain rights and that one man had been killed as a result of a wound from a poisoned arrow and one young girl

[37] NAN, KAB 1, W. Volkmann, 30 Oct. 1928, Report on the Agricultural and Political Conditions at the Angola Boundary.

[38] Kreike, *Re-creating Eden*, chap. 7.

[39] NAN, NAO 36, District Surgeon Ondangwa, Monthly Reports Aug.-Sep. 1927.

[40] NAN, NAO 17, Eedes to Hahn, Namakunde, 9 April 1928.

wounded".[41] In 1930, San from Ongandjera in western Ovamboland northwest of Etosha Pan attacked migrant laborers from Uukwaluthi district.[42]

The perception of the region's San as living in a state of upheaval is also punctuated by evidence of a crisis in terms of livelihoods. The late 1920s and early 1930s were drought years that triggered famine in the Ovambo floodplain and led to San behavior that colonial officials described as being "very rare". In August 1927, for example, seven San men from Angola applied at the recruitment center at Ondangwa for work as migrant laborers. In September 1927, another four "Ondonga-Bushmen" followed suit. Two of the seven San recruits from Angola were diagnosed as having light cases of scurvy, suggesting nutritional stress.[43] It is unclear whether the colonial officials were surprised because they had never heard of San men engaging in migrant labor, or because the San whom they perceived to lead a simple but idyllic existence were forced to engage with 'civilization'.[44]

How could the San, who were an integral part of the 'wilderness', be so dramatically affected by a subsistence crisis that they resorted to theft and wage labor? Colonial reports from the 1930s and 1940s explicitly distinguished between the 'wild Bushmen' who were nomadic hunter-gatherers and the 'tame Bushmen' who were being absorbed within Ovambo culture. The 1,300–1,600 San who lived east of the Ovambo floodplain largely were considered to be as 'wild' as the Kalahari San and they were perceived to "lead a carefree, though at times in drought seasons, a hard life". The 1928 annual report for Ovamboland mentioned that the eastern Ovamboland San "when they feel the pinch of hunger, move towards the Ovambo settlements, where they generally manage to barter a little grain or other food stuffs".[45] A 1948 report stated that the San suffered from venereal disease and "habitual coughs" with a high death rate, but added that "[they] seem happy and care free".[46] In contrast, the several hundred San in western Ovamboland, whom

[41] NAN, NAO 11, O / C NAO to Clarke, Ondangwa, 6 June 1929. See also NAO 18, Annual Report Ovamboland 1928.

[42] NAN, NAO 18, Monthly Reports Ovamboland, May – June 1930.

[43] NAN, NAO 36, District Surgeon Ondangwa, Monthly Reports Aug.-Sep. 1927.

[44] Lee argues that the San and hunter-gatherers in general are much less vulnerable to subsistence crises than agriculturalists because their main food consists of wild vegetables and fruits which he considers a "stable" resource, Lee, "What Hunters Do for a Living", pp. 30–43.

[45] NAN, NAO 18, Annual Report Ovamboland 1928. The 1934 annual report also emphasizes that San acquired grain, beans and tobacco, NAO 19.

[46] NAN, NAO 61, Annual Report Ovamboland 1948.

the colonial administration had forced to resettle closer to the floodplain in order to facilitate colonial control, engaged in agriculture and raised livestock including goats, sheep, cattle and donkeys, and they were consequently no longer considered 'wild'.[47]

If hard times indeed had fallen upon the San as a result of the decline of regional game populations, violence, social dislocation and physical displacement, then it is all the more surprising that their 1920s and 1930s home environment was deemed a 'wilderness' despite being neither pristine nor 'natural'. Although the 'wilderness' east of the floodplain and south of the Angolan-Namibian border does not contain any known remnants of ancient cities or densely populated centers, it was to a significant extent a domesticated environment. San communities constructed the water holes that formed the basis for their settlements, some of which were permanent. Digging and maintaining water holes was extremely labor-intensive; excess sand and litter had to be cleared away every year. Indeed, the Assistant Native Commissioner of Ovamboland, Bourquin, distinguished between San *camps* which were temporary, and San *villages*: in late May 1940, his party failed to locate two San camps and was forced to spend the night at Omboto, where he discovered a "large permanent Bushmen settlement", which he identified as "the Bushmen village of old Ule, the leader of the group of that area".[48] In a subsequent report, however, although Ule is referred to as a "clan" leader whose authority extended over several San groups in the Omaheke, suggesting that they were organized in a chiefdom, the inhabitants of the area were thought to have been "untouched by civilisation and must be considered as typical wild bushmen living in their natural environment".[49] The San water holes provided critical stepping stones for the Ovambo floodplain migrants who expanded settlement into the eastern Ovamboland wilderness.[50] For example, in 1941 the Assistant Native Commissioner reported:

[47] NAN, NAO 20, Annual Reports Ovamboland 1938, and 1942; NAO 61, Annual Report Ovamboland 1948.

[48] NAN, NAO 10, ANC to NCO, Oshikango, 30 July 1940, Report on Development Work undertaken in Eastern Ukuanyama during 1940, and NAO 20, Monthly Report Ovamboland, May 1940.

[49] NAN, NAO 20, Annual Report Ovamboland 1942.

[50] A 1928 sketch map indicates eight locations with water holes between Ohinengena (Ohangwena; just east of the inhabited zone of Oukwanyama at the time) and Otufima (Outafima), including one location in Angola. A San settlement is explicitly indicated at Nehova Pan. Some of these

> When we reached Ombongolo on 29/8/41 we found that the Ukuanyamas had cleaned and enlarged the Bushman waterhole at the spot (...). at Okulukila a dry Bushman waterhole was found but unfortunately time did not allow us to have it cleaned (...). At Shikome (...) two small Bushman waterholes about six feet deep were found in use. These were cleaned and enlarged and gave a good supply of water (...). Further shallow Bushman waterholes were found at Omboto and Oshiti and there is no doubt that it will be possible to sink many waterholes in this locality.[51]

In the same year, another report mentioned Kules or Kures [Kudis?], forty to fifty miles northwest of Tsintsabis Police Station, where there were more or less permanent water holes, "and where there are always Bushmen. The present leader is Nainabab". Yet another report talked about a San "Chief" called Hamketeb in the Omaheke.[52]

San communities in eastern Ovamboland temporarily abandoned their settlements for the duration of the dry season in order to follow migrating wildlife. In 1942, the Assistant Native Commissioner reported:

> the Omaheke bush is bushmen country. For generations Bushmen have lived here and have established more or less permanent settlements. Each group has its well defined territory ant [*sic*] it moves from waterhole to waterhole within its boundary.[53]

Despite the observation that San settlements were permanent and despite the explicit association of important water holes with San communities, colonial officials nevertheless considered these water holes to be open-access natural resources

---

water holes may have been San water holes and/or water holes dug by herdsmen. See map accompanying NAN, NAO 17, Acting UGR to O/C NAO Namakunde, 25 Feb. 1928. On a sketch map dating from 1932, a San water hole and settlement are indicated at Ombongola, south of beacon 40. Another San water hole is indicated between beacons 37 and 38, and a third near Outafima between beacons 29 and 30. See NAO 17, Officer Commanding Oshikango (McHugh), Oshikango, March [date illegible], 1932.

[51] NAN, NAO 10, ANC to NCO, Oshikango, 6 Sep. 1941, Proposed Extension of Ukuanyama up to and including Ombongolo Muramba.

[52] NAN, NAO 21, NCO to Deputy Commissioner South African Police, Windhoek, 31 Dec. 1941, and Station Commander SAP to District Commander SAP Omaruru, Tsintsabis, 20 Feb. 1942.

[53] The ANC added that "[n]o reliable information could be obtained from the local Bushmen who do not appear to travel far afield but remain in more or less well defined areas in the Omaheke country". NAN, NAO 10, ANC to NCO, Oshikango, 10 July 1942, Proposed Extension of Ukuanyama Area: General Report Development Work Eastern Ukuanayama.

and colonial officials and ethnographers alike maintained that ownership of environmental resources was an alien concept within San culture.[54] In the Ovambo concept of water ownership, a water hole was either owned by the person who dug it, by his or her descendants, or by the person who maintained it. The Ovambo conceptualization meshed with the views of colonial officials, who believed the San were at best 'underutilizing' the land, while the Ovambo agropastoralists were considered to be able to use the land much more efficiently. The subsequent colonization of the region east of the Ovambo floodplain by Angolan refugees and migrants from the floodplain, legitimized by Ovambo and colonial officials' (re)definition of the San as 'wild' and their environment as 'wilderness', thus set in motion a process that marginalized the San as environmental, economic and social actors. San groups and individuals lost control over much of the eastern 'wilderness' and its resources. For example, Matias Kafita, a San, hunted and gathered with his father and other San during his youth but as an adult was merely relegated to assisting horse-mounted Ovambo hunters.[55]

The structural erosion of San livelihoods by the 1950s became obvious even to colonial officials. Whereas the 1948 annual report for Ovamboland had distinguished 'tame Bushmen' and 'wild Bushmen', the 1951 annual report only used the term 'semi-tame Bushmen' to categorize 2,500 people in the census that year.[56] During the 1952 drought, Ovamboland's San were the primary recipients of free food aid, while in 1953, the eastern San temporarily received government assistance, including medical aid at what was effectively a famine refugee camp at Omundaunghilo.[57] At the same time, officials recruited the few San remaining in the Namutoni (Etosha) Game Reserve in a project to collect animal bones. The scheme was initiated in 1950 to allow the San to earn some cash and to supply the bone meal factory at Okahandja. San bone collectors

---

[54] NAN, NAO 10, ANC to NCO, Oshikango, 8 Oct. 1942. Lee acknowledges that the Dobe-San of the Botswana Kalahari Desert were not truly nomadic and notes that main San camps were located at a permanent water hole, Lee, "What Hunters Do for a Living", in pp. 31, 35.

[55] Matias Kafita, interview by author, Ekoka laKula, 23 Feb. 1993. See also Moses Kakoto, interview by author, Okongo, 17 Feb. 1993.

[56] NAN, NAO 61, Annual Reports Ovamboland 1948 and 1951.

[57] NAN, NAO 60, Quarterly Report Ovamboland, 4th Quarter 1952, and NAO 64f. 19/1, Medical Officer Ovamboland, Monthly Reports Aug.-Sep. 1953.

received 2s. 6d. per 100 pounds. By the end of 1950, almost 50,000 pounds had been purchased.[58] In 1954, however, all San residents in the park were expelled.[59]

During the 1950s, the San remaining on both sides of the border east of the floodplain received food aid and tobacco as incentives to assemble at the stations of the Finnish Mission, which in turn resettled them at Okongo and Ekoka in mission villages where they were trained in crop cultivation.[60] A missionary took Matias Kafita and his family from their Eenhana home first to Okongo and subsequently to Ekoka. When they arrived at Ekoka,

> there was another tribe of San called Ovakwagga !Kung ([they] call themselves Kwagga). My community [was] called Ovagongolo (…). The Kwagga were like us also in the forest and were hunting animals. The Kwagga left when the Mission [the FMS] came here. At that time [there] were no wild animals around. They gave us the garden; it was for us not for the Mission. We got enough food from the big garden and maize (…). At the beginning the garden was cultivated as one garden for everybody. When the missionaries found out that we could do the work properly they divided the garden among the people. [There was] only one Kwanyama household headed by Johannes Namidi. He was the village headman of Ekoka. He was looking after the San and was given that task by Eliki [Erik, an FMS missionary].[61]

It was during this stage in the history of the San in eastern Ovamboland that they became the focus of a surge in academic interest. During 1951, for example, the Portuguese ethnographer Martin Gusinde engaged in research on the region's San and a research group led by Laurence Marshall visited the area. The government's Bushman Commission conducted research in Ovamboland during 1952.[62] By that time, however, the alienation of the San's environmental resource base in the central Angolan-Namibian border region had been justified and the San marginalized in the name of colonial development and modernization.

---

58 NAN, NAO 59, NCO to Sec. SWA, Ondangwa, 8 Dec. 1950, and NCO to Station Commander SAP Namutoni, Ondangwa, 14 Jan. 1950; Station Commander SAP Namutoni to NCO, Namutoni, 20 Dec. 1950 and Director Agriculture to Sec. SWA, Windhoek, 22 Oct. 1952.

59 NAN, NAO 61, Annual Report Ovamboland 1953.

60 Interviews by author: Matias Kafita, Ekoka laKula, 23 Feb. 1993 and Moses Kakoto, Okongo, 17 Feb. 1993.

61 Matias Kafita, interview by author, Ekoka laKula, 23 Feb. 1993.

62 NAN, NAO 60, Quarterly Reports Oukwanyama, April – June and July-Sep. 1951. On the Bushman Commission, NAO 106 Diary NCO 1949–1952, entries 30 July, 2 and 3 Aug. 1952.

That entire ecosystems and individual species, including humans, historically can vacillate between the categories 'wild' and 'domesticated' challenges the Nature-Culture dichotomy. Not only were the San Naturalized, but the same was true for the floodplain's human and domestic and wild animal populations and indeed its entire environment. And colonialism imposed the Naturalization: after a late precolonial era of profound engagement in global and regional markets (ivory, iron, copper) in the late nineteenth century. By the mid-twentieth century, however, the region's San had become vagabonds surviving on food aid. The experience is identical to the deglobalizing, localizing and Naturalizing process that affected Ovambo floodplain cattle owners but, ultimately, it was much more destructive to the San in its economic and social consequences. The situation is all the more ironic because in a Nature-Culture dichotomy, colonialism is the major agent and mechanism for dislodging 'non-Western' societies and environments from the grip of Nature, and launching them toward development on a Nature-to-Culture path. While the modernization paradigm qualifies this pathway as being for the best, the declinist and inclinist paradigms judge it as being for the worse. The environmental history of Ovamboland sheds grave doubts on the validity of any paradigm that conceptualizes environmental change as a unilinear and irreversible path from a state of pristine wilderness / Nature to a state of domestication / Culture with the urban landscape at the apex. Such a conceptualization results in a paradox whereby environments defined as 'wilderness' have been marked and shaped by intensive human use in a deeper or more recent past, as was the case, for example, at Palenque. And the Palenques are only the tip of the iceberg: the semi-urban and rural humanized environments that sustained similar urban centers must have stretched over much of the modern forest and savanna wilderness landscapes of Central America and Africa, including such an 'African Eden' as the Kruger National Park in South Africa.

8

The Ovambo paradox and environmental pluralism

The modernization, declinist and inclinist paradigms portray environmental change as unilinear, irreversible and homogenous. As the previous chapters demonstrate, the paradigms are unilinear because they describe change in linear fashion that occurs along a Nature-to-Culture (or wilderness-humanized landscape) gradient. Depending on the paradigm, change is progressive, for the better or for the worse, as well as cumulative, and often irreversible. The paradigms homogenize analysis because they depict environmental change as an undifferentiated single process with a singular outcome: degradation, improvement or a stable state.

Descriptions of the early twentieth-century Ovambo floodplain environment are strikingly similar to those of the late twentieth century: both depict settlements characterized by farms neighboring each other, with towering fruit trees and dense woody vegetation located on the edges of the farms and between the villages. At first glance, the similarities suggest a stable state or environmentally sustainable land use. An alternative interpretation might be that any changes, for the worse or the better, essentially canceled each other out over the course of the twentieth century. Yet, population pressure, the colonial project, biological invaders and technological and economic changes caused dramatic environmental transformations throughout the last century.

Before World War II, tens of thousands of refugees fled from the northern floodplain villages into the wildernesses of the middle floodplain and beyond. From the 1920s onward, following the end of a long era of violence and insecurity, southern floodplain settlers fanned out from their fortified and congested villages into the surrounding landscapes, where they came into contact with the refugees from the north, as well as with recovering wildlife populations. The northern floodplain refugees and the southern floodplain migrants deforested the land to construct new farms, fields and eventually new villages. The South African colonial administration encouraged the colonization of the wilderness which was shaped by the creation of virtual and actual temporary or permanent inter- and intracolonial boundaries that limited the movement of people, animals and goods, especially beyond the new borders of the Ovamboland reserve. The boundaries were imposed where necessary through displays of military power as evidenced, for example, by orders to shoot animals on sight to prevent the spread of lungsickness and foot and mouth following the establishment of veterinary cordons. The result was a reconcentration of local, refugee or 'invader' human populations. Deprived of firearms, the refugee-settlers were exposed to new diseases and supplied with new tools (notably the plow) in a shrinking physical space that intensified concerns about environmental degradation. These concerns in turn shaped the political, social and moral ecologies of the floodplain.

The human and animal populations in Ovamboland increased overall, but inconsistently, and with staggering contradictions. Until the 1950s, rough estimates indicated that cattle populations fluctuated enormously from year to year. Interannual variation was at least in part probably a reflection of a combination of high take-off rates and the impact of disease and drought. The 1980s witnessed a decline in cattle numbers. The number of donkeys, a species introduced in the floodplain early in the century, languished for several decades, only to increase rapidly after World War II, mainly as a consequence of their import into Ovamboland. While gun technology and guns were adopted rapidly in the floodplain after their introduction in the second half of the nineteenth century, imported metal hoes and plows did not have much of an impact until after World War II. From the 1950s onward, however, the adoption of the plow had a strong direct and indirect environmental impact. Overall, the impact of population pressure, green and biological imperialisms, the market economy and Western technology in Ovamboland was neither linear nor straightforward.

The reforestation that followed in the wake of deforestation is a dramatic illustration of this point. As refugees and migrants transformed their new environments, they propagated pre-existing and new woody vegetation. When perceived in cyclical terms, the occurrence of successive deforestation and reforestation is neither unique nor new. The waxing and waning of forests characterized, for example, Ghana (forest clearing between 1000 and 1600 and again in the 1900s), the Ethiopian Highlands, the *miombo* bush savannas of eastern and southern Africa (with expansion and contraction spanning at least the last 22,000 years) and the forests of the Midwestern and Eastern United States (where oak forests have repeatedly expanded and contracted during the last 10,000 years).[1]

But this was not simply cyclical change, as in a return to a (vegetation) climax. In the United States, for example, Native American use of fire fostered a forest dominated by such fire-resistant species as oak, hickory and chestnut. When the use of fire was suppressed, such fire-sensitive species as red maple and sugar maple replaced oak; moreover, forest encroached and continues to advance on what used to be savanna or barrens. The composition of the Central American forests of today is also dissimilar to that of the forests that marked the pre-Mayan environment. Furthermore, Japan saw massive reforestation in the wake of World War II, but two-thirds of its mountain forests are industrial monoculture forests.[2] Processes of afforestation that do not directly result from human agency, as occurs in forest plantations, for example, but rather from natural regrowth, as in the case

[1] On Ghana, Fairhead and Leach, *Reframing Deforestation*, pp. 76–77. On Ethiopia, McCann, *Green Land, Brown Land*, pp. 79–107. On the *miombo*, see Campbell, *The Miombo in Transition*, pp. 5–6, and Misana, Mung'ong'o and Mukamuri, "Miombo Woodlands in the Wider Context". On the United States, see McShea and Healy, "Oaks and Acorns as a Foundation for Ecosystem Management", pp. 4–5; and Abrams, "The Postglacial History of Oak Forests in Eastern North America". See also Williams, *Deforesting the Earth*, pp. 3–4, 12.

[2] On the United States, see Abrams, "The Postglacial History of Oak Forests in Eastern North America"; and D. Dey, "Fire History and Postsettlement Disturbance" and "The Ecological Basis for Oak Silviculture in Eastern North America". See also Brown, Curtin and Brathwaite, "Management of the Semi-Natural Matrix", pp. 331–336. On Namibia, see Kreike, *Re-creating Eden*, chap. 8, and "Architects of Nature", pp. 57–62. On Central America, see Leyden, Brenner, Whitmore, Curtis, Piperno and Dahlin, "A Record of Long- and Short-Term Variation from Northwest Yucatán: Cenote San José Culchacá". On Japan, see Knight, "From Timber to Tourism", pp. 335–336.

of the re-establishment of forests and woodlands on abandoned lands or as the result of fire suppression, also highlight Nature's role as an actor rather than a victim of or backdrop to human initiative.[3]

Thus, whereas many areas were heavily deforested and reforested, the process of environmental change was neither cyclical nor neatly sequential, but rather highly ambiguous in terms of its baseline, dynamics and outcome. While acknowledging ambiguity in the record of environmental change is not new, such ambiguity may be attributed to different valuations of what constitutes degradation and of what may be considered to be improvement. Moreover, interpretations of the significance of the process of environmental change and its outcome may differ.[4] For example, a case study of a former ranch in southern Chiapas in Mexico from the 1960s through the 1990s identified a "*simultaneous* recovery of degraded forest lands and intensification of maize cultivation".[5] Both afforestation and environmental decline occurred in Wällo, Ethiopia.[6] In 1950s to 1980s northern Ivory Coast, a decline in wildlife (unambiguous degradation) coincided with a simultaneous increase of cropland and woodland (or afforestation) at the expense of open bush land.[7] A baseline survey of 1,800 households in a Zimbabwean afforestation

---

[3] See for example Cronon, *Changes in the Land*. McCann points out that the role of climate is relatively understudied in the recent environmental historiography of Africa, see "Climate and Causation in African History".

[4] Moore and Vaughan, *Cutting Down Trees*; and Fairhead and Leach, *Misreading the African Landscape*. Blaikie and Brookfield stress that a multidisciplinary approach and a sensitivity to multicausality are required for environmental study but regard environmental change as monoprocessual, in *Land Degradation and Society*, pp. 14–16. Meggers attributes the conflicting interpretations about environmental change in the Amazon to a lack of communication between scholars from different disciplinary backgrounds, in "Natural Versus Anthropogenic Sources of Amazonian Biodiversity", p. 89. See also Mazzucato and Niemeijer, *Rethinking Soil and Water Conservation in a Changing*, pp. 114–116; and Gibson, McKean and Ostrom, "Explaining Deforestation", p. 2. Schama notes that the impact of humans on Nature is not an unmixed blessing, but he stresses that the ways humans affect Nature do not constitute "unrelieved and predetermined calamity" either. See Schama, *Landscape and Memory*, pp. 9–10.

[5] Van der Haar, "Peasant Control and the Greening of the Tojolabal Highlands, Mexico", pp. 110–112. Emphasis added.

[6] Crummey and Winter-Nelson, "Farmer Tree Planting in Wällo, Ethiopia", esp. p. 119. See McCann, *People of the Plow*.

[7] Bassett, Koli Bi and Okattara, "Fire in the Savanna", esp. p. 64. On Zimbabwe, see Kerkhof, *Agroforestry in Africa*, pp. 69–73. A. Erkkilä notes both deforestation and regrowth in woody vegetation cover in some areas, in "Living on the Land", pp. 73–75, 99–101. Gibson et al. attribute the disagreement that exists concerning the underlying causes of deforestation to the possibility that multiple processes are at work, to knowledge gaps or to both, Gibson, McKean and Ostrom, "Explaining Deforestation", p. 2.

project revealed that deforestation was strongly correlated with clearing land for crop cultivation, but that the non-arable land was not deforested and might in fact have gained woody biomass.[8]

In the example of the former ranch in Chiapas, the ranch was construed as a land use system which served as the spatial unit of analysis. The paradoxical findings of intensification of agriculture and afforestation, however, may have been partly an artifact of the relatively abstract scale of the analysis, because the study lacked specific data to illuminate the step-by-step processes of environmental change.[9] Scale of analysis is a critical variable for analyzing the process of environmental change and for evaluating its outcome. Larger-scale outcomes average out outcomes at smaller scales. For example, on a global scale, the second half of the twentieth century witnessed severe deforestation, but the United States and Western Europe actually experienced reforestation.[10] Twentieth-century Bangladeshi farmers planted trees on homestead mounds but at the same time cleared trees in the surrounding floodplain to make fields.[11] If the homestead mound gardens were the unit of analysis, the outcome of the process of environmental change would have been afforestation; if, on the other hand, the actual floodplain were the focus, the assessment would have been one of deforestation. If the Bangladeshi floodplain land use system as a whole were to be evaluated, the outcome would depend on the amount of afforestation on the mounds and the extent of deforestation in the plain. The scale of analysis thus may significantly influence the outcome of the assessment. Multiscale analysis may partially counter this issue; as Huxley notes, however, "research activities are nearly always confined to a single scale level".[12]

---

[8] Kerkhof, *Agroforestry in Africa*, pp. 69–73.

[9] Van der Haar, "Peasant Control and the Greening of the Tojolabal Highlands, Mexico", esp. pp. 110–112.

[10] Williams, *Deforesting the Earth*, pp. 412–431.

[11] Leuschner and Khaleque, "Homestead Agroforestry in Bangladesh".

[12] Huxley, *Tropical Agroforestry*, p. 302. On multiscale analysis, see Gibson, McKean and Ostrom, "Explaining Deforestation". Gibson et al. note contradictions in environmental trends and the need to differentiate environmental change. G. Varughese, for example, studied eighteen villages in the middle hills of Nepal and found radically different environmental trends: in seven villages the forest was degrading, in six villages it was improving and in five the forest conditions were stable, Varughese, "Population and Forest Dynamics in the Hills of Nepal", esp. p. 204, table 8.2. Gibson and Becker noted enormous variation in how individuals in Western Ecuador used their forest plots: some clear-cut their plots and others encouraged regrowth, Gibson and Becker, "A Lack of Institutional Demand", esp. pp. 135–136, 156.

Alternatively, the ambiguity in the record of environmental change may be a reflection of the plural nature of the process of environmental change. Different readings of the process and outcome of environmental change are not merely misreadings or divergent interpretations of a singular, homogenous process of environmental change from various points of view, inspired by, for example, the modernization, declinist or inclinist paradigm, or by a population pressure, political ecology or biological imperialism approach. Rather, the differences may result from the need to analyze environmental changes as multiple (sub)processes of environmental changes—occurring at different scales and even within a single scale—that, moreover, are neither synchronized, nor discrete, nor fully interlinked.

Deforestation in Ovamboland

Ovamboland in the twentieth century saw both deforestation and reforestation. Wood use was neither naturally sustainable, nor did reforestation gains simply cancel out deforestation losses over the long run, an outcome that might suggest equilibrium and the absence of fundamental environmental change. The exact opposite was true: environmental change was dramatic. Deforestation in Ovamboland from the 1920s through the 1950s was only too real. As people fled south across the colonial border from the northern floodplain or fanned out from the heartlands of the occupied southern floodplain polities, they cut woody vegetation to construct homesteads and to clear fields. For example, Native Commissioner Hahn wrote in 1931:

> The Ovambos, who are agriculturalists, when they established themselves in the first instance, cut away into the bush and cleared spaces to make room for their fields. The timber and scrub thus cut away is firstly used to build their pallisaded [*sic*] kraal and secondly to enclose the borders of their lands, etc.[13]

Officials and missionaries, especially in the 1920s and 1930s, witnessed a conjuncture of deforestation as thousands upon thousands of refugees and migrants

[13] NAN, SWAA 3, NCO to Sec. SWA, Ondangwa, 20 April 1931.

streamed into the wilderness areas of Ovamboland. The specter must have appeared similar to the massive forest clearing that threatens the forests of, for example the Amazon and Indonesia today. In Ovamboland's wilderness, settlers not only cut down large amounts of poles to construct huts, palisades, kraals and fences, but also obliterated most of the tree and bush vegetation on their prospective farm plots. Fire was the settlers' most powerful tool when clearing land for farms and fields. A concerned agricultural expert commented in 1924:

> Natives are very destructive of the natural bush & their method of clearing ground is not economical. The usual method is to put a fire around a tree until it falls, no effort being made to remove the stump (...). The destruction of the bush, without any effort to replant in suitable places will mean at an early date the extension of the desert & it is a problem requiring immediate & careful attention.[14]

Assuming that the average farm size was between 0.5 and 2 ha, the approximately 18,000 households that were counted in the 1933 census meant that over time, between 9,000 and 36,000 ha had been cleared of woody vegetation in order to make room for homesteads, kraals and crop fields. But since much of the wood for the actual construction of the farms had been obtained elsewhere than on the farms themselves, the creation of 18,000 farms theoretically would have led to the deforestation not only of the 9,000–36,000 ha of farm plots but also of an estimated 9–18,000 ha of *mopane* bush land required for construction materials, for a grand total of approximately 18–54,000 ha of affected bush land or 'forest'. Deforestation was most dramatic in Ovamboland's Oukwanyama district, which was located directly south of the Angolan border, where the approximately 6,000 new farms that were established between 1916 and 1933 consumed between 3,000 and 6,000 ha of *mopane* bush land and resulted in the further clearing of 3,000–12,000 ha of farm plots. Deforestation thus affected 6,000–18,000 ha in a seventeen-year

[14] NAN, NAO 26 Report Ovamboland Cotton Prospects, appendix to Alec Crosby to Bishop of Damaraland [Ms.], St. Mary's Mission, 11 Jan. 1924. See also O/C NAO Oshikango to NCO, Oshikango, 20 June 1938; NAO 10, ANC to NCO, 31 Oct. 1940; Hahn's handwritten notes on the letter "Also in regard to indiscriminate burning of Mopane trees in Ukuambi and Ukuanyama lands"; SWAA 3 f. Forestry: Indigenous Forests Ovambo A1/2 (I), NCO to CNC, Ondangwa, 2 June 1941; BAC 131, Agricultural Officer Ovamboland to Bantu Commissioners Ondangwa and Oshikango [Ondangwa?], 28 Jan. 1957.

Photo 3. New Homestead, c. 1928. A large number of poles cut from the surrounding area were used to construct the palisades and the frames and walls of huts, leaving the fields bare. This large newly constructed homestead where fruit trees have not established themselves yet may have belonged to the Oukwanyama senior headman Nyukuma Shingelifa. Encouraged by Native Commissioner Hahn, the headman settled at Omhedi in the middle floodplain wilderness in 1927 or 1928. Hahn showcased Nyukuma's homestead to dignitaries and other visitors (National Archives of Namibia, Hahn Collection)

period, for an average of 353–1,059 ha affected per year.[15] Moreover, most of this dramatic deforestation took place in a relatively small area in the middle floodplain. In the second half of the 1920s, the impact was especially concentrated in the area directly west, south and east of Oshikango along the border, right under the nose of the Assistant Native Commissioner of Ovamboland who was based there. The dramatic nature of deforestation was also heightened by the fact that all the new farms were located on the low ridges in between the watercourses, with farm plots cleared on the lower slopes and construction wood obtained from the upper slopes. The watercourses themselves could not be used for habitation or cultivation because they flooded.[16]

[15] On wood consumption for constructing homesteads, see Kreike, "Architects of Nature", pp. 96–98 and 141–146.

[16] NAN, NAO 104, Anderson to Hahn, diary Jordan, and A233, J. Chapman, 1903–1916, 1876[?],

As the population of Ovamboland grew from 107,000 in 1933 to 126,000 in 1938, then to 197,000 in 1951 and 618,000 in 1991, the amount of land that was cleared for fields commensurately increased.[17] Based on a small survey sample, a 1991 report estimated the average farm size to range from 2 to 5 ha, with farms in eastern Ovamboland being larger on average than in the actual floodplain.[18] Thus, with 90,918 'traditional' rural homesteads having been counted in 1991, an estimated 181,836 to 454,590 ha of the total area of Ovamboland's 4,200,000 ha, i.e., from 4.3 to 10.8 %, had been transformed into farm plots, compared to 9,000–36,000 ha or 0.21–0.86 % in 1933. The colonial administration estimated aggregate farmland in Ovamboland to be 27,606 ha in 1950, 71,961 ha in 1957, 59,968 ha in 1958, 88,400 ha in 1966, 94,000 ha in 1968 and 150,000 ha in 1978–1979. The figures for 1957 forward all seem to have been estimates of the actual surface area that was being cultivated, rather than of the total available farm area, and they therefore underestimate the total land cleared.[19]

---

pp. 61–62; Kreike, *Re-creating Eden*, chap. 2; Lima, *A Campanha dos Cuamatos*, pp. 132–114; AVEM, RMG 2518 [?] C/h 52, Speiker, Visitationsbericht, Namakunde 13–18 July 1906; AGC-SSp, Duparquet, journal, no. 6, 1878, 1881, information from Carlston; Petrus Shanika Hipetwa, interview by author, Oshiteyatemo, 17 June 1993. NAN, KAB 1 (iii), W. Volkmann, 30 Oct. 1928, "Report on the Agricultural and Political Conditions at The Angola Boundary". See also AHE (BAC) 1/346, Report of the SWA Planning Committee for Agricultural Training Centers, appendix to Chief Bantu Affairs Commissioner SWA to Bantu Affairs Commissioners Ondangwa, Runtu and Oshikango, [Windhoek], 8 April 1965. This situation is also borne out by regular reports of flooded fields and destroyed crops, see, for example NAN, NAO 18, Monthly Reports Ovamboland, March and April 1925. See also NAO 19, Monthly Reports Ovamboland, Feb., June and July 1934, Jan. and Feb. 1937, and NAO 21, Quarterly Report Ovamboland, Jan. – March 1944. Similar conditions marked 1950, NAO 60, Quarterly Report Ovamboland, Jan. – March and April – June 1950.

[17] NAN, OVA 53, Sec. SWA to Sec. Agriculture Owambo, Windhoek, 24 June 1974, Appendices A–C.

[18] Namibian Economic Policy Research Unit, "Land related Issues in the Communal Areas, 1: Owambo" (Windhoek: Paper for the National Land Conference, 1991). The author's personal observations (19 Feb. 1993) bear out the impression that especially in the far east, in such villages as, for example, Ehafo, Oshikuni and Big and Little Olukula, farm plots were considerably larger than in the floodplain. The plots were usually fenced with wire or branches and there were very few trees in the fields.

[19] Figures provided in morgen are given in hectares. See NAN, NAO 103, Census of Agriculture Ovamboland 1949–1950; BAC 133, Agricultural Report Ovamboland 1956–1957 and Quarterly Report Agriculture Ovamboland for the Quarter ending 30 June 1958; AHE (BAC) 1/352 [1/357], Annual Reports Agriculture Ovamboland, 1966 and 1968; OVA 6, Annual Report Agriculture Ovamboland 1978–1980.

The creation of new and larger farms and fields, facilitated by plow technology, led to heavy and often dramatic deforestation. On-farm deforestation was more destructive than cutting construction wood off-farm because most woody vegetation was burned, a method of clearing that limited regrowth.

Reforestation in Ovamboland

Paradoxically, as deforestation of the wilderness areas of Ovamboland progressed from the 1920s through the 1940s, a process of reforestation followed in its wake. What is most striking about reforestation in Ovamboland is that the farms and fields that were the locus of the most destructive deforestation simultaneously were the areas where the most spectacular reforestation occurred. In a matter of decades, majestic fruit trees towered over homesteads and crops. Moreover, reforestation was not confined to the farms or to fruit trees. Although wood harvesting had been especially severe in the 1920s and 1930s when refugees and migrants streamed into the wilderness area along the Angola-Namibian colonial border, clearing farm plots and constructing palisaded homesteads, deforestation beyond the actual farm plots had been somewhat less destructive: poles and branches had been removed without killing the plants, and many woody vegetation species in Ovamboland had the ability to resprout. In addition, the original woodlands of new villages partially were retained, albeit in the form of a heavily managed bush coppice.[20]

Although the striking presence of on-farm fruit trees figures prominently in colonial and postcolonial descriptions of Ovamboland's vegetation, colonial officials and experts and their postcolonial successors found it difficult to imagine that the existence of the fruit trees constituted reforestation. Despite noting that fruit trees occurred on farms and fields, and that they were mostly absent from wilderness areas, colonial officials do not seem to have been shaken in their belief that, for example, marula and birdplum were wild trees. But marula (*Sclerocarya birrea*) and birdplum (*Berchemia discolor*) trees, and to a lesser extent the real fan palm (*Hyphaene petersiana*), by and large only began to appear in the middle

20 Kreike, "Architects of Nature", pp. 150–180, 208–219.

floodplain wilderness after the area was settled during the 1920s and 1930s. Moreover, although they sometimes occurred randomly, the phenomenon was confined to a highly specific and private space: farms and fields, where they were protected by palisades, fences and / or herdsmen from people, livestock and the elements.

Oral histories confirm that the introduction of marula and birdplum trees in the middle floodplain wilderness accompanied settlement. In early 1930s Omupanda, marula trees were confined to only two of the eight existing farms, and one of the two was located on the farm of the first person who had settled there around 1900. In the Ombalantu and Uukwambi districts, many marula and birdplum trees had likewise been planted by the households who had constructed new villages in the uninhabited stretches that had separated the precolonial polities.[21]

Paulus Wanakashimba attributed the introduction of marula and birdplum trees in his village to the women who had collected the fruit in older villages further north; some of the pits that had been discarded after the fruit had been consumed had developed into seedlings. Paulus Nandenga, however, emphasized that careful human management facilitated the 'natural' propagation of fruit trees: "[Seedlings] only survived because during the dry season [they] lose their leaves and animals cannot eat them. During the rainy season, if they are located in the fields, people will till the soil and prevent the goats from entering".[22] Indeed, although Paulus Wanakashimba's village had few marula trees when he was a young boy during the early 1920s, by the mid 1930s, both his and the neighboring village boasted many marula and birdplum trees. After clearing his own farm in 1947, he fenced new seedlings with thorn bush to protect them from livestock, and by the early 1990s, his farm contained birdplum, marula and real fan palm trees.[23]

---

[21] Interviews by author: Mateus Nangobe, Omupanda, 24 May 1993; Paulus Wanakashimba, Odimbo, 10–11 Feb. 1993; Paulus Nandenga, Oshomukwiyu, 28 April 1993; Kulaumoni Haifeke, Oshomukwiyu, 11 May 1993. On Uukwambi, personal communication from Joseph Hailwa, Regional Forester Ovamboland, 24 March 1992. On the occurrence of marula and birdplum in late 1960s Ombalantu, see NAN, OVA 57, Lueckhoff, Report on Visit to SWA, 3–15 Nov. 1969, appendix to Regional Forester to Chief Director Bantu Administration, Grootfontein, 3 April 1970.

[22] Paulus Nandenga, interview by author, Oshomukwiyu, 28 April 1993.

[23] Paulus Wanakashimba, interview by author, Odimbo, 10–11 Feb. 1993.

Photo 4. Mature Homestead and Fields, 1993. The homestead is partly hidden behind the trees to the left. This is an old farm in the southern floodplain, witness the large trees (including real fan palm and marula) that tower over the fields (Photo by author)

Beginning in the 1930s, a fruit tree frontier advanced beyond the Ovambo floodplain into eastern Ovamboland. Omundaunghilo, east of the floodplain, was already a fully fledged village by 1923 with birdplum, marula and real fan palm trees. But in most of the region, mature floodplain fruit trees appeared later because settlement only really took off during the early 1920s. A 1934 report on settlement in the east stated: "The usual fruit trees are, of course, not as plentiful as in the actual tribal area but natives are being encouraged to plant them whenever possible".[24] Kalolina Naholo observed settlers in the east seeding marula, birdplum and jackalberry (*Diospyros mespiliformis*). Marula could also be propagated by cutting off a green branch and planting it in moist soil.[25]

[24] NAN, A450, 7, Annual Report 1935.

[25] Interviews by author: Kalolina Naholo, Ohamwaala, 26–27 Jan. 1993; Paulus Nandenga, Oshomukwiyu, 28 April 1993; Franscina Herman, Odibo, 12 Dec. 1992; cf. Helemiah Hamutenya, Omuulu Weembaxu, 17 July 1993. On eastern Ovamboland, see Kreike, *Re-creating Eden*, chap. 9.

While the floodplain filled with farms, fields and fruit trees during the 1950s–1990s, the fruit tree frontier advanced into far eastern Ovamboland, towards the border with Okavango.[26] In 1993, small birdplum trees could be found as far east as Olukula. Beyond Olukula, however, birdplum, marula and real palm trees were rare.[27] Moses Kakoto settled in Okongo on an existing farm during the late 1960s. Although the birdplum trees on his farm had grown 'naturally', he had planted palm seeds from the floodplain in his first homestead.[28] Timotheus Nakale stressed that fruit trees were more numerous in fields in the west, i.e., the floodplain, because they grew "naturally"; in the east, however, people had had to plant the seeds. During the 1960s and 1980s, he had planted marula and birdplum seeds that he had brought from further west, and these grew into large trees. In 1992, when he moved his homestead to a new location, he successfully seeded more birdplum in addition to palm seeds; he had obtained the latter from Uukwambi. Some of the fruit trees, notably jackalberry, had not grown at all.[29]

People actively and passively propagated fruit trees on-farm because the surrounding fenced farm plots and the palisaded homesteads were the locations where they prepared, consumed and prepared fruit, and where they discarded the seeds, for example on middens and through defecation, or planted seeds and cuttings. Fences and palisades kept browsing goats, cattle and donkeys at bay and provided shade. The inhabitants discarded used water and other waste on-farm, and organic matter from decaying poles and other construction materials enriched the soil.[30]

The twentieth-century history of environmental change in the Ovambo floodplain constitutes an enormous interpretive challenge if it is approached within a binary deforestation / reforestation framework. The modernization, declinist and

---

[26] Tree propagation in the older villages continued. See, for example, Philippus Haidima, interview by author, Odibo, 9 Dec. 1992.

[27] Werner Nghionanye, interview by author, Olukula laKula, 18 Feb. 1993, and personal observations, 20 Feb. 1993. In a survey of 35 crop fields in the west of eastern Ovamboland, Erkkilä found that marula, birdplum and real fan palm trees occurred with the highest frequency, and that these trees only occurred near or on crop fields, see Erkkilä, "Living on the Land", pp. 96–97.

[28] Interviews by author: Kalolina Naholo, Ohamwaala, 26 and 27 Jan. 1993; Kaulikalelwa Oshitina Muhonghwo, Ondaanya, 2 Feb. 1993; Moses Kakoto, Okongo, 17 Feb. 1993.

[29] Timotheus Nakale, interview by author, Ekoka laKula, 21 Feb. 1993.

[30] Kreike, "Architects of Nature", pp. 157–180.

inclinist paradigms frame change in terms of a singular process with a singular outcome: either environmental degradation or improvement. The paradox in Ovamboland is that environmental change is characterized by a process of environmental degradation in the form of deforestation that simultaneously is accompanied by a process of environmental recovery in the form of reforestation, creating a paradox.

Environmental pluralism: Multiprocessual asynchronous environmental change

Colonial officials and experts 'misread' Ovamboland's landscape. By presuming that the majestic fruit trees that marked Ovamboland were wild and wilderness trees, they failed to understand the full extent to which the local inhabitants had shaped the environment, through both deforestation and reforestation. The officials and experts also described a population that was entirely subjected to Nature and too 'primitive' to embrace the blessings of colonial development, causing overgrazing and desertification. But colonial claims about environmental degradation were grossly overdrawn. Moreover, colonial actions had in fact (re)Naturalized aspects of local land use. Colonial policies resulted in the deglobalization of pastoralism and hunting and the decommodification of domestic and wild animals, while conservation and disarmament led to a resurgence of wild animal predation.

Still, colonial assessments of the region's environment were not entirely misreadings: the observations of massive deforestation in the 1920s and 1940s, for example, were to a significant degree correct. Thus, exclusively reading colonial records against the grain might be counterproductive. The crux of the matter is to accept the ambiguities for what they are, rather than attempting to average them out, and to be alert to the shortcomings of the paradigms that dominate the analysis, conceptualization and representation of environmental change.

The process of environmental change cannot, in fact, be understood solely in terms of a linear Nature-to-Culture (or wild-to-domestic) dichotomy. This argument has been made by, amongst others, William Cronon and Richard White, but it needs to be made more explicit, because the dichotomy lies at the core of the modernization, declinist and inclinist paradigms. The paradigms, their concepts and their vocabulary by definition re-introduce the Nature-Culture dichotomy into the analysis and description. Moreover, as conceptualized here, the process of environmental change is not necessarily singular, homogenous, synchronous, self-contained, or even coherent. Rather, environmental dynamics inherently are complex and plural, consisting of multiple strands, trajectories and subprocesses that may converge and diverge in asynchronous asymmetry.

Pathways of environmental change may be contradictory: the record of deforestation and reforestation is but one, though perhaps the most dramatic, example of contradictory trajectories of environmental change in twentieth-century Ovamboland. Human population pressures (first population redistribution and subsequently natural increase) triggered and shaped both the nature and the extent of deforestation and reforestation. At the same time, cattle population pressures had no appreciable *direct* effect on the environment, even while colonial officials cried wolf about overgrazing and desertification. The *indirect* impact, however, was significant because colonial anxieties about overgrazing, desertification and livestock diseases fueled green imperialism and led to tremendous changes in livestock use and management in the floodplain. Green imperialism in Ovamboland in turn was far from homogenous, despite a high level of continuity amongst its main protagonists. Two successive Native Commissioners (Hahn and Eedes) ruled the colonial administration of Ovamboland for three and a half decades. Both men championed wildlife conservation while openly refusing to implement or enforce most of the other conservation measures that were standard practice in South Africa's Native Reserves. Eedes practiced green imperialism, but he also undercut it by successfully deflecting challenges to his personal and autocratic rule by a new generation of scientifically trained colonial agricultural and forestry experts.

The impact of the modern market economy in Ovamboland was equally diffuse. After the imposition of colonial rule, floodplain men were more tightly sucked into a regional (and global) labor market, especially from the 1940s onward, but at the same time floodplain society was cut off from global livestock, game, slave and small arms markets. Such biological invaders as lungsickness and rinderpest had a devastating virgin soil epidemic impact (even as Africa supposedly shared the

Old World disease environment with Europe), but foot and mouth did not, even though it served to legitimize the draconian intensification of green imperialism. Finally, Western technology offers a mixed record: guns were rapidly adopted and then disappeared almost as fast following disarmament, leading respectively to a quick decline of large game populations in the floodplain, and to its subsequent resurgence.

The subprocesses of environmental change often were also asynchronous and their outcomes fleeting. Floodplain wildlife and livestock populations in the early 1900s were at an unrepresentative low as a result of rinderpest and the global demand for ivory and cattle. Wildlife populations increased until the 1940s, even as the population of Ovamboland soared due to in-migration, and then declined. The floodplain populations rapidly adopted guns as soon as they were introduced in the late 1800s, but imported metal tools and implements only replaced locally produced agricultural tools from the late 1940s, although they had been introduced at least two decades earlier. Lungsickness, rinderpest, measles, influenza and plague had an immediate impact, but the effect of another biological invader—the donkey—was minimal until the late 1940s. The subsequent proliferation of donkeys was less a product of the resetting of an innate biological switch than of its newfound utility as a draft and transport animal.

Analysis based on the modernization, declinist and inclinist paradigms does not fully capture these and other intricacies of environmental change. When used as absolute benchmarks, such State of Nature-derived concepts as (natural) climax (vegetation) and (natural) biodiversity obscure as much as they reveal. The wilderness, the wildlife and the 'wild Bushmen' in Ovamboland were not in a state of nature in the immediate precolonial era. In fact, in the late 1890s, the wilderness of Ovamboland barely hid the ruins of a once prosperous kingdom, hunting and disease had reduced its wildlife populations to a low, and the local San engaged in commercial hunting. Present-day states of Nature (or states of Culture) are equally problematic benchmarks. Upon closer examination, presumed relic islands of natural vegetation may be far from pristine. Using the present as a absolute benchmark 'state of Culture' outcome along a Nature-to-Culture gradient of environmental change is also unacceptably teleological because it treats the present as the only or the necessary outcome of past processes that in fact may not fully have run their course.

An alternative method for assessing environmental change is to establish a series of empirically derived analytical benchmarks.[31] These multiple and historical benchmarks are by definition relational, not absolute. As demonstrated by the case of Ovamboland, comparing a 1890s state of the environment with a 1990s state may at first glance suggest that little or no change has occurred. The apparent continuity between the two moments in time, however, conceals dramatic vegetation change: deforestation and reforestation in the time frame of less than a century. Only the use of multiple measuring points and a focus on the processes of change facilitates detecting environmental changes. Such change can be gradual but it can also occur quite rapidly: the 1896–1897 rinderpest epizootic, for example, decimated susceptible wild and domestic animal populations across the African continent.

The presence of human settlement ruins in pristine nature, the evidence that wild plants have been planted, and simultaneous deforestation and reforestation constitute contradictions only if environmental change is conceived as a unilinear, irreversible and singular process with a singular outcome within the framework of a Nature-Culture dichotomy that has Nature as the point of departure. Rather, environmental change should be imagined as a series of subprocesses that can be asymmetric and asynchronous. Differentiating the process of environmental change requires a more open-ended research focus: for example, assess environmental change(s) rather than measure deforestation or reforestation. Rejecting Nature in its various incarnations of pristine Nature, relic Nature, (biological) climax and biodiversity, as a useful benchmark to study environmental change does not entail an aesthetic rejection of Nature per se. It also does not mean the death of Nature. Furthermore, it does not mean that everything is Culture in the sense that humanity dominates Nature and that previously natural environments have been irreversibly polluted (in a declinist perspective) or improved (in a modernization perspective).

In addition, treating ecosystems like closed and independent systems can result in framing ecosystemic change in terms of a zero-sum game, in which one organism's gain is by definition another's loss because the total resources within

[31] For examples of such an approach, see Harms, *Games against Nature*; McCann, *People of the Plow*; and Knapen, *Forests of Fortune?*

the system are finite. External dynamics are as critical in evolutionary terms as internal dynamics. Energy flows are the most dramatic example: ecosystem Earth is energized by solar radiation either directly, through photosynthesis, climate and the weather, or indirectly, through fossil fuels as stored solar energy. Yet, humanity's capacity optimally to utilize the terrestrial and extraterrestrial resources available without destroying itself and the organic and inorganic forms with which it shares the earth is a product of the historical context of any particular point in time. Whereas the modernization paradigm inspires unbridled optimism in this respect, the declinist paradigm engenders morbid pessimism. The inclinist paradigm offers a more guarded optimism. All three paradigms, however, present the Nature-Culture struggle as a zero-sum game: natural resources are limited and careful stewardship is seen to be essential. In brief, in the declinist paradigm, good stewardship is seen to be lacking (resulting in degradation); in the modernization paradigm, good stewardship is defined as (Western) scientific stewardship; and in the inclinist paradigm, good stewardship is defined as indigenous stewardship.

Evolution as a process is a product of internal and external dynamics, the dynamics and the fostering conditions of which are time-specific; they are historic (including in the wider meaning of geological time-depth), and thus by definition subject to change. Thus, in every moment in time, the evolutionary process displays an array of entities and a series of interactions between entities that cannot be seen only as (fleeting) outcomes of the process, and the genetic and behavioral precursors of the next phase of the process, but which are also unique to that moment in time (and decisively so if an entity were to become extinct).

In addition, each historical moment is marked by a unique biodiversity. This means that even if no (pristine) Nature is left, each moment in history is valuable not only in purely historical terms, but also in terms of biodiversity and genetic diversity. Thus, historicizing Nature and Culture highlights the need to preserve the earth's bionic inheritance not only for its historical and aesthetic significance, but also because preserving a variety of historical landscapes, ecosystems and organisms provides access to historical biodiversity and gene pools. Moreover, there is no convincing evidence that biodiversity was significantly higher in a 'State of Nature' (if it ever existed) than it is today; rather, evolutionary theory and genetics suggest the opposite: organisms and genes became more diverse over time. Yet, even if the former were true and bio- and genetic diversity have narrowed between a hypothetical 'State of Nature' and the present, for example, through ecocide, the cumulative bio- and genetic diversity produced throughout history would be infinitely larger than that of any single moment in time, whether it be the era of pristine Nature or the present. Even if much of the sum of history's bio- and genetic diversity probably has been lost, today's ecosystem Earth and its bio- and genetic diversity are not the static outcome of one genealogical line of evolution. Rather, ecosystem Earth at all its levels reflects a huge variety of more or less related processes of environmental changes that are sometimes synchronized, and sometimes not. In fact, in the context of evolution-as-history, a static outcome may result in an evolutionary dead end and extinction.

This does not mean that concern about the state of the environment is unwarranted; to the contrary. What is preserved because it is thought to represent Nature or the (more) Natural is not less pleasing aesthetically because it actually may not be as pristine as was presumed. Moreover, environments that have been rejected as unworthy of careful management and preservation because they were seen to be all Culture, or, worse, as neither Culture nor Nature, may need to be integrated in biopreservation. Select environments are singled out for preservation as Cultural Heritage; however, the environments that are neither fully Nature nor fully Culture typically are neglected, even though they have their own unique richness of bio- and genetic diversity. How to preserve them is a challenge: applying conventional conservation formulae that severely limit human use and management of these environments is socially, politically, economically and logistically impossible. Moreover, eliminating or proscribing the actions of one set of human agents (and substituting them by another: the conservation experts) from those environments would change the environmental dynamics, and cause environmental change as a result of the very intervention that is meant to preserve the environment. Instead, analyses of environmental change and extinction should seek to go beyond conceptualizations of Nature-to-Culture and homogenous processes and outcomes and focus on understanding environmental pluralism as a means for addressing the environmental challenges that confront Earth, both in Africa and elsewhere on the planet.

Bibliography

Abrams, M.D. "The Postglacial History of Oak Forests in Eastern North America". McShea and Healy, *Oak Forest Ecosystems*, pp. 34–45.

Adams, J.S. and T.O. McShane. *The Myth of Wild Africa: Conservation without Illusion*. Berkeley: University of California Press, 1996 [first published 1992].

Adjei, M.B. and J.P. Muir. "Current Developments from Tropical Forage Research in Africa". Sotomayor-Ríos and Pitman, *Tropical Forest Plants*, pp. 331–355.

"Ainda o Desastre do Humbe". *Portugal em África* 5(52) (March 1898): 55.

Akyeampong, E.K. *Between the Sea and the Lagoon: An Eco-Social History of the Anlo of Southeastern Ghana: C. 1850 to Recent Times*. Athens: Ohio University Press, 2001.

Alcorn, J.B. "Huastec Noncrop Resource Management: Implications for Prehistoric Rain Forest Management". *Human Ecology* 9(4) (1981): 395–417.

Amanor, K.S. *The New Frontier: Farmer Responses to Land Degradation: A West African Study*. Geneva: UNRISD, 1994.

Amaral Ferreira, R. "Transforming Atlantic Slaving: Trade, Warfare and Territorial Control in Angola, 1650–1800". PhD thesis, University of California, Los Angeles, 2003.

Anderson, D. "Managing the Forest: The Conservation History of Lembus, Kenya, 1904–63." Anderson and Grove, *Conservation in Africa*, pp. 249–268.

Anderson, D. "Depression, Dust Bowl, Demography, and Drought: The Colonial State and Soil Conservation in East Africa during the 1930s". G. Maddox, ed. *Colonialism and Nationalism in Africa*, vol. 2: *The Colonial Epoch in Africa*. New York: Garland, 1993, pp. 209–231.

Anderson, D. and R. Grove, eds. *Conservation in Africa: People, Policies and Practice*. Cambridge: Cambridge University Press, 1987.

Aperçue historique: Chronique des missions confieés à la congrégation du Saint Esprit, 1930–1931. Paris: Maison Mère, 1932.

Arnold, D. *The Problem of Nature: Environment, Culture, and European Expansion*. Oxford: Blackwell, 1996.

Audric, J. *Angkor Wat and the Khmer Empire*. London: R. Hale, 1972.

Balée, W. "Indigenous History and Amazonian Biodiversity". Steen and Tucker, *Changing Tropical Forests*, pp. 185–197.

Bassett, T.J., Z. Koli Bi and T. Okattara. "Fire in the Savanna: Environmental Change and Land Reform in Northern Côte d'Ivoire". Bassett and Crummey, *African Savannas*, pp. 53–71.

Bassett, T.J., and D.E. Crummey, eds. *Land in African Agrarian Systems*. Madison: University of Wisconsin Press, 1993.

Bassett, T.J. and D. Crummey, eds. *African Savannas: Global Narratives and Local Knowledge of Environmental Change*. Oxford: James Currey; Portsmouth, N.H.: Heinemann, 2003.

Beach, D.N. *The Shona and Zimbabwe, 900–1850*. Gweru: Mambo Press, 1990 [first published 1980].

Becker, C.D. and R. León. "Indigenous Forest Management in the Bolivian Amazon: Lessons from the Yuracaré People". Gibson, McKean and Ostrom, *People and Forests*, pp. 163–191.

Beinart, W. "Soil Erosion, Conservationism, and Ideas about Development: A Southern African Exploration, 1900–1960". *Journal of Southern African Studies* 11(1) (1984): 52–83.

Beinart, W. "Introduction: The Politics of Colonial Conservation". *Journal of Southern African Studies* 15(2) (1989): 143–162.

Beinart, W. "Soil Erosion, Animals, and Pasture over the Longer Term: Environmental Destruction in Southern Africa". Leach and Mearns, *The Lie of the Land*, pp. 54–72.

Beinart, W. *The Rise of Conservation in South Africa: Settlers, Livestock, and the Environment, 1770–1950*. Oxford: Oxford University Press, 2003.

Beinart, W. and C. Bundy. *Hidden Struggles in Rural South Africa: Politics and Popular Movements in the Transkei and Eastern Cape, 1890–1930*. London: James Currey, 1987.

Beinart, W., P. Delius and S. Trapido, eds. *Putting a Plough to the Ground: Accumulation and Dispossession in Rural South Africa, 1850–1930*. Johannesburg: Ravan Press, 1986.

Beinroth, F.H. "Land Resources for Forage in the Tropics". Sotomayor-Ríos and Pitman, *Tropical Forest Plants*, pp. 3–15.

Berry, S. *Cocoa, Custom, and Socio-Economic Change in Rural Western Nigeria*. Oxford: Clarendon Press, 1975.

Berry, S. "Social Institutions and Access to Resources". *Africa* 59(1) (1989): 41–55.

Berry, S. *No Condition Is Permanent: The Social Dynamics of Agrarian Change in Sub-Saharan Africa*. Madison: University of Wisconsin Press, 1993.

Bertelsmann, W. "Wasserbau im Ovamboland". *SWA Annual*, 1959, pp. 141–144.

Blaikie, P. and H. Brookfield, with contributions by B. Allen et al. *Land Degradation and Society*. London: Methuen, 1987.

Bonnéhin, L. "Domestication paysanne des arbres fruitiers forestiers: Cas de Coula edulis Baill. olacaceae, et de Tieghemella heckelii Pierre ex. A. Chev., sapotaceae, autour du Parc National de Tai, Côte d'Ivoire". PhD diss., Wageningen Agricultural University, 2000.

Boomgaard, P. "Exploitation and Management of the Surinam Forests, 1600–1975". Steen and Tucker, *Changing Tropical Forests*, pp. 252–264.

Boomgaard, P. *Frontiers of Fear: Tigers and People in the Malay World, 1600–1950*. New Haven: Yale University Press, 2001.

Boserup, E. *The Conditions of Agricultural Growth: The Economics of Agrarian Change under Population Pressure*. New York: Aldine, 1965.

Boyce, S.G. *Landscape Forestry*. New York: Wiley, 1995.

Bradshaw, G.A. and P.A. Marquet, eds. *How Landscapes Change: Human Disturbance and Ecosystem Fragmentation in the Americas*. Berlin: Springer Verlag, 2003.

Brooks, G.E. *Landlords and Strangers: Ecology, Society, and Trade in West Africa, 1000–1630*. Boulder, Colo.: Westview Press, 1993.

Brown, J.H., C.G. Curtin and R.W. Brathwaite. "Management of the Semi-Natural Matrix". Bradshaw and Marquet, *How Landscapes Change*, pp. 327–344.

Budowski, G. "Perceptions of Deforestation in Tropical America: The Last 50 Years". Steen and Tucker, *Changing Tropical Forests*, pp. 1–4.

Bundy, C. *Rise and Fall of the South African Peasantry*. London: Heinemann, 1979.

"Campanha do Cuamato". *Portugal em Africa* 14(165) (September 1907): 443–448.

Campbell, B., ed. *The Miombo in Transition: Woodlands and Welfare in Africa*. Bogor, Indonesia: CIFOR, 1996.

Campbell, B.M. "The Use of Wild Fruits in Zimbabwe". *Economic Botany* 41(3) (1987): 375–385.

Capello, H. and R. Ivens. *De Angola a Contra-Costa: Descripção de uma Viagem atravez do Continente Africano*. Lisbon: Impr. Nacional, 1886.

Carruthers, J. *The Kruger Park: A Social and Political History*. Pietermaritzburg: University of Natal Press, 1995.

Carson, R. *Silent Spring*. New York: Houghton Mifflin, 1994 [first published 1962].

Chandler, D.P. *A History of Cambodia*. Boulder, Colo.: Westview Press, 1992.

Chew, S.C. *World Ecological Degradation: Accumulation, Urbanization, and Deforestation, 3000 BC–AD 2000*. Walnut Creek, Calif.: AltaMira Press, 2001.

Chidumayo, E., J. Gambiza and I. Grundy. "Managing Miombo Woodlands". Campbell, *The Miombo in Transition*, pp. 175–193.

Clarence-Smith, W.G. *Slaves, Peasants, and Capitalists in Southern Angola 1840–1926*. Cambridge: Cambridge University Press, 1979.

Cleaver, K.M. and G.A. Schreiber. *Reversing the Spiral: The Population, Agriculture, and Environment Nexus in Sub-Saharan Africa*. Washington, D.C.: World Bank, 1994.

Cohen, D.W. and E.S. Atieno Odhiambo. *Siaya: A Historical Anthropology of an African Landscape*. London: James Currey, 1989.

Colchester, M. "Colonizing the Rainforests: The Agents and Causes of Deforestation". Colchester and Lohmann, *The Struggle for Land*, pp. 1–15.

Colchester, M. "Forest Peoples and Sustainability". Colchester and Lohmann, *The Struggle for Land*, pp. 61–95.

Colchester, M. and L. Lohmann, eds. *The Struggle for Land and the Fate of the Forests*. Penang, Malaysia: the World Rainforest Movement, 1995 [reprint; first published 1993].

Coley, P.D. and T.A. Kursar. "Anti-Herbivore Defenses of Young Tropical Leaves: Physiological Constraints and Ecological Trade-Offs". S.S. Mulkey, R.L. Chazdon and A.P. Smith, eds., *Tropical Forest Plant Ecophysiology*. New York: Chapman & Hall, 1996, pp. 305–336.

Cooper, F. and A.L. Stoler, eds. *Tensions of Empire: Colonial Cultures in a Bourgeois World.* Berkeley: University of California Press, 1997.

Cooper, G.J. *The Science of the Struggle for Existence.* Cambridge: Cambridge University Press, 2003.

Cordell, D.D. and J.W. Gregory, eds. *African Population and Capitalism: Historical Perspectives.* Boulder, Colo.: Westview Press, 1987.

Coucher, C.L. and E.W. Herbert. "The Blooms of Banjeli: Technology and Gender in West African Iron Making". Schmidt, *The Culture and Technology of African Iron Production*, pp. 40–57.

Cox, C.B. and P.D. Moore. *Biogeography: An Ecological and Evolutionary Approach.* 6th ed. Oxford: Blackwell, 2000 [first published 1973].

Craddock, P.T. *Early Metal Mining and Production.* Edinburgh: Edinburgh University Press, 1995.

Cronon, W. *Changes in the Land: Indians, Colonists, and the Ecology of New England.* New York: Hill and Wang, 1988 [reprint; first published 1983].

Cronon, W. *Nature's Metropolis: Chicago and the Great West.* New York: Norton, 1991.

Crosby, A.W. *The Columbian Exchange: The Biological and Cultural Consequences of 1492.* Westport, Conn.: Greenwood, 1972.

Crosby, A.W. *Ecological Imperialism: The Biological Expansion of Europe, 900–1900.* Cambridge: Cambridge University Press, 1986.

Crummey, D. and A. Winter-Nelson. "Farmer Tree Planting in Wällo, Ethiopia". Bassett and Crummey, *African Savannas*, pp. 91–120.

Curtin, P.D. *Disease and Empire: The Health of European Troops in the Conquest of Africa.* Cambridge: Cambridge University Press. 1998.

Dalal-Clayton, B. "Environmental Aspects of the Bangladesh Flood Action Plan". International Institute for Environment and Development (IIED), Issues Series No. 1 (1990).

Darrow, K. "Provenance-type Trials of Eucalyptus Camaldulensis and Eucalyptus Tereticornis in South Africa and South-West Africa: Eight-year Results". *South African Forestry Journal* 124 (March 1983): 13–22.

Davidson, P. "Museums and the Reshaping of Memory". Nuttall and Coetzee, *Negotiating the Past*, pp. 143–160.

Dean, W. *With Broadax and Firebrand: The Destruction of the Brazilian Atlantic Forest.* Berkeley: University of California Press, 1997.

Dejene, A. *Environment, Famine, and Politics in Ethiopia: A View from the Village.* Boulder, Colo.: Lynn Rienner, 1990.

Denevan, W.M. *Cultivated Landscapes of Native Amazonia and the Andes.* Oxford: Oxford University Press, 2002.

Denning, G.L. "Realising the Potential of Agroforestry: Integrating Research and Development to Achieve Greater Impact". Franzel, Cooper, Denning and Eade, *Development and Agroforestry*, pp. 1–14.

De Wet, C. "Betterment Planning in a Rural Village in Keiskommahoek, Ciskei". *Journal of Southern African Studies* 15(2) (1989): 326–345.

Dey, D. "The Ecological Basis for Oak Silviculture in Eastern North America". McShea and Healy, *Oak Forest Ecosystems*, pp. 60–79.

Dey, D. "Fire History and Postsettlement Disturbance". McShea and Healy, *Oak Forest Ecosystems*, pp. 46–59.

Diamond, J. *Guns, Germs, and Steel: The Fates of Human Societies*. New York: W.W. Norton & Company, 1999 [first published 1997].

Diamond, J. *Collapse: How Societies Choose to Fail or Succeed*. New York: Viking, 2005.

Doornbos, M., A. South and B. White, eds. *Forests: Nature, People, Power*. Oxford: Blackwell, 2000.

Drinkwater, M. "Technical Development and Peasant Impoverishment: Land Use Policy in Zimbabwe's Midlands Province". *Journal of Southern African Studies* 15(2) (1989): 287–305.

Drummond, J. and A. Manson. "The Rise and Demise of Agricultural Production in Dinokana Village, Bophuthatswana". *Canadian Journal of African Studies* 27(3) (1993): 462–479.

Dunham, P.S. "Resource Exploitation and Exchange among the Classic Maya". Fedick, *The Managed Mosaic*, pp. 315–332.

Dunlap, T. *Nature and the English Diaspora: Environment and History in the United States, Canada, Australia, and New Zealand*. Cambridge: Cambridge University Press, 1999.

Dunning, N.P. "A Reexamination of Regional Variability in the Pre-Hispanic Agricultural Landscape". Fedick, *The Managed Mosaic*, pp. 53–91.

Durham, W. *Scarcity and Survival in Central America: Ecological Origins of the Soccer War*. Stanford, Calif.: Stanford University Press, 1979.

Ehrlich, P. *The Population Bomb*. New York: Ballantine, 1968.

Ehrlich, P.H. and A.H. Ehrlich. *The Population Explosion*. New York: Simon & Schuster, 1990.

Eldredge, E.A. *A South African Kingdom: The Pursuit of Security in Nineteenth Century Lesotho*. Cambridge: Cambridge University Press, 1993.

Emmett, T. *Popular Resistance and the Roots of Nationalism in Namibia, 1915–1966*. Basel: Schlettwein, 1999.

Erkkilä, A. and H. Siiskonen. *Forestry in Namibia, 1850–1990*. Joensuu: University of Joensuu, 1992.

Erkkilä, A. "Living on the Land: Change in Forest Cover in North-Central Namibia, 1943–1996". PhD diss., University of Joensuu, 2001.

Estermann, C. "Notas Ethnográficas sobre os Povos Indígenas do Distrito da Huíla", *Boletim Geral das Colónias* 11(116) (Feb. 1935): 4–69.

Estermann, C. *Etnografia do Sudoeste de Angola*, vol. 1: *Os Povos não-Bantos e o Grupo Étnico dos Ambós*. [Lisbon]: Ministério do Ultramar, Junta de Investicações do Ultramar, 1960.

Evans, J.G. *Environmental Archaeology: Principles and Methods*. Phoenix Mill, UK: Sutton Publishing, 1999.

Fairhead, J. and M. Leach. *Misreading the African Landscape: Society and Ecology in a Forest-Savanna Mosaic*. Cambridge: Cambridge University Press, 1996.

Fairhead, J. and M. Leach. *Reframing Deforestation: Global Analysis and Local Realities. Studies in West Africa*. London: Routledge, 1998.

Farley, J. *Bilharzia: A History of Imperial Tropical Medicine*. Cambridge: Cambridge University Press, 1991.

Fedick, S.L., ed. *The Managed Mosaic: Ancient Maya Agriculture and Resource Use*. Salt Lake City: University of Utah Press, 1996.

Feierman, S. *Peasant Intellectuals: Anthropology and History in Tanzania*. Madison: University of Wisconsin Press, 1990.

Fenn, E. *Pox Americana: The Great Smallpox Epidemic of 1775–1782*. New York: Hill & Wang, 2001.

Ferguson, J. *The Anti-Politics Machine: "Development", Depoliticization, and Bureaucratic Power in Lesotho*. Minneapolis: University of Minneapolis, 1994.

Ferreira Diniz, J. de Oliveira. *Negocios Indigenas: Relatório do Ano de 1913*. Loanda: Imprensa Nácional, [1913].

First, R. *Black Gold: The Mozambican Miner, Proletarian and Peasant*. New York: St. Martin's Press, 1983.

Ford, J. *The Role of Trypanosomiases in African Ecology*. Oxford: Clarendon Press, 1971.

Fortmann, L. "The Tree Tenure Factor in Agroforestry with Particular Reference to Africa". *Agroforestry Ecosystems* 2(4) (1985): 229–251.

Foster, L.V. *A Brief History of Central America*. New York: Facts on File, 2000.

Franzel, S. et al. "Methods of Assessing Agroforestry Adoption Potential". Franzel and Scherr, *Trees on the Farm*, pp. 11–36.

Franzel, S. and J. Scherr, eds. *Trees on the Farm: Assessing the Adoption Potential of Agroforestry Practices in Africa*. Wallingford, UK: CABI, 2002.

Franzel, S., P. Cooper, G.L. Denning and D. Eade, eds. *Development and Agroforestry: Scaling Up the Impacts of Research*. Oxford: Oxfam, 2002.

Frédéric, L. *La vie quotidienne dans la Péninsule indochinoise à l'Epoque d'Angkor*, 800–1300. Paris: Hachette, 1981.

Frost, P. "The Ecology of Miombo Woodlands". Campbell, *The Miombo in Transition*, pp. 11–57.

Gade, D.W. *Nature and Culture in the Andes*. Madison: University of Wisconsin Press, 1999.

Geertz, C. *Agricultural Involution: The Processes of Environmental Change in Indonesia*. Berkeley: University of California Press, 1963.

Giblin, J. "Trypanosomiasis Control in African History: An Evaded Issue?" *Journal of African History* 31(1) (1990): 59–80.

Giblin, J. *The Politics of Environmental Control in Northeastern Tanzania, 1840–1940*. Philadelphia: University of Pennsylvania Press, 1992.

Giblin, J. "The Precolonial Politics of Disease Control in the Lowlands of Northeastern Tanzania". Maddox, Giblin and Kimambo, *Custodians of the Land*, pp. 127–151.

Gibson, C.C. and C.D. Becker. "A Lack of Institutional Demand: Why a Strong Local Community in West Ecuador Fails to Protect Its Forest". Gibson, McKean, and Ostrom, *People and Forests*, pp. 135–161.

Gibson, C.C., M.A. McKean and E. Ostrom. "Explaining Deforestation: The Role of Local Institutions". Gibson, McKean and Ostrom, *People and Forests*, pp. 1–26.

Gibson, C.C., M.A. McKean and E. Ostrom, eds. *People and Forests: Communities, Institutions, and Governance*. Cambridge, Mass.: MIT Press, 2000.

Gillet, H. "Observations on the Causes of Devastation of Ligneous Plants in the Sahel and Their Resistance to Destruction". Le Houérou, *Browse in Africa*, pp. 127–129.

Goody, J. *Technology, Tradition, and the State in Africa*. London: Oxford University Press, 1971.

Gordon, R.J. *The Bushman Myth: The Making of a Namibian Underclass*. Boulder, Colo.: Westview Press, 1992.

Goudsblom, J, *Fire and Civilisation*. London: Penguin, 1992.

Graham, E. and D.M. Prendergast. "Maya Urbanism and Ecological Change". Steen and Tucker, *Changing Tropical Forests*, pp. 102–109.

Gray, L.C. "Investing in Soil Quality: Farmer Responses to Land Scarcity in Southwestern Burkina Faso". Bassett and Crummey, *African Savannas*, pp. 72–90.

Grinde, D. and B. Johansen. *Ecocide of Native America: Environmental Destruction of Indian Lands and Peoples*. Santa Fe, N.Mex.: Clear Light, 1995.

Grove, R. "Colonial Conservation, Ecological Hegemony and Popular Resistance: Towards a Global Synthesis" in MacKenzie, *Imperialism and the Natural World*, pp. 15–50.

Grove, R.H. *Green Imperialism: Colonial Expansion, Tropical Island Edens and the Origins of Environmentalism 1600–1860*. Cambridge: Cambridge University Press, 1997 [first published 1995].

Guha, R. *The Unquiet Woods: Ecological Change and Peasant Resistance in the Himalaya*. Berkeley: University of California Press, 1989.

Guy, J. "Ecological Factors in the Rise of Shaka and the Zulu Kingdom". S. Marks and A. Atmore, eds. *Economy and Society in Pre-Industrial South Africa*. London: Longman, 1980, pp. 102–119.

Haggar, J. et al. "Participatory Design of Agroforestry Systems: Developing Farmer Participatory Research Methods in Mexico". Franzel, Cooper, Denning and Eade, *Development and Agroforestry*, pp. 15–23.

Hall, M. *The Changing Past: Farmers, Kings, and Traders in Southern Africa, 200–1860*. Cape Town: D. Philip, 1987.

Hallé, F., R.A.A. Oldeman and P.B. Tomlinson. *Tropical Trees and Forests: An Architectural Analysis*. Berlin: Springer Verlag, 1978.

Hanna, W.W. and S. Torres-Cardona. "*Pennisetums* and *Sorghums* in an Integrated Feeding System in the Tropics". Sotomayor-Ríos and Pitman, *Tropical Forest Plants*, pp. 193–200.

Harms, R. *Games against Nature: An Eco-cultural History of the Nunu of Equatorial Africa*. Cambridge: Cambridge University Press, 1987.

Harper, K.T., G.B. Ruyle and L.R. Rittenhouse. "Toxicity Problems Associated with the Grazing of Oak in Intermountain and Southwestern USA". James, Ralphs and Nielsen, *The Ecology and Economic Impact of Poisonous Plants on Livestock Production*, pp. 197–206.

Harries, P. *Work, Culture, and Identity: Migrant Laborers in Mozambique and South Africa, c. 1860–1910*. Portsmouth, N.H.: Heinemann, 1994.

Harrison, P.D. and B.L. Turner II, eds. *Pre-Hispanic Maya Agriculture*. Albuquerque: Univerity of New Mexico Press, 1978.

Hartland, W.A. "Maya Settlement Patterns: A Critical Review". E. Wyllys Andrews IV et al., eds., *Archaeological Studies in Middle America*. New Orleans: Middle American Research Institute, Tulane University, 1970, pp. 21–47.

Hartwig, G.W. and K.D. Patterson, eds. *Disease in African History: An Introductory Survey and Case Studies*. Durham, N.C.: Duke University Press, 1978.

Hayes, P. "'Cocky' Hahn and the 'Black Venus': The Making of a Native Commissioner in South West Africa, 1915–1946". N.R. Hunt et al., eds., *Gendered Colonialisms in African History*. Oxford: Blackwell, 1997, pp. 42–70.

Hayes, P., J. Silvester, M. Wallace and W. Hartmann, eds. *Namibia under South African Rule: Mobility and Containment, 1915–1946*. Oxford: James Currey, 1998.

Headrick, D.R. *The Tools of Empire: Technology and European Imperialism in the Nineteenth Century*. New York: Oxford University Press, 1981.

Headrick, R. *Colonialism, Health and Illness in French Equatorial Africa, 1885–1932*. Edited by D.R. Headrick. Atlanta, Georgia: ASA Press, 1994.

Hendricks, F.T. "Loose Planning and Rapid Resettlement: The Politics of Conservation and Control in Transkei, 1950–1970". *Journal of Southern African Studies* 15(2) (1989): 306–325.

Henkemann, A.B., G.A. Persoon and F.K. Wiersum. "Landscape Transformations of Pioneer Shifting Cultivators at the Forest Fringe". Wiersum, *Tropical Forest Resource Dynamics and Conservation*, pp. 53–69.

Herbert, Eugenia W. *Iron, Gender, and Power: Rituals of Transformation in African Societies*. Bloomington: Indiana University Press, 1993.

Herskovits, M.J. "The Cattle Complex in East Africa". PhD thesis, Columbia University, 1926.

Higham, C. *The Archaeology of Mainland Southeast Asia from 10,000 BC to the Fall of Angkor Wat*. Cambridge: Cambridge University Press, 1989.

Hobley, M. *Participatory Forestry: The Process of Change in India and Nepal*. London: ODI, 1996.

Hoffmann, R. "La importancia ecológica y económica de las tecnologías tradicionales en la agri- y silvicultura en áreas de bosque tropical humido en México". Steen and Tucker, *Changing Tropical Forests*, pp. 110–122.

Holland, A. "Ecological Integrity and the Darwinian Paradigm". Pimentel, Westra and Noss, *Ecological Integrity*, pp. 45–59.

Huxley, P. *Tropical Agroforestry*. Oxford: Blackwell Science, 1999.

Iliffe, J. *A Modern History of Tanganyika*. Cambridge: Cambridge University Press, 1984 [first published 1979].

Iliffe, J. *Africa: The History of a Continent*. Cambridge: Cambridge University Press, 1995.

Isaacman, A. *Cotton Is the Mother of Poverty: Peasants, Work, and Rural Struggle in Colonial Mozambique, 1938–1961*. Portsmouth: N.H.: Heinemann, 1996.

Isaacman, A. and R. Roberts, eds. *Cotton, Colonialism, and Social History in Sub-Saharan Africa*. Portsmouth, N.H.: Heinemann, 1995.

Isaacman, A. and C. Sneddon. "Toward a Social and Environmental History of Cahora Bassa Dam". *Journal of Southern African History* 26(4) (2000): 597–632.

Isenberg, A.C. *The Destruction of the Bison*. Cambridge: Cambridge University Press, 2000.

Jacobs, N.J. *Environment, Power and Injustice: A South African History*. Cambridge: Cambridge University Press, 2003.

James, L.F., M.H. Ralphs and D.B. Nielsen, eds. *The Ecology and Economic Impact of Poisonous Plants on Livestock Production*. Boulder, Colo.: Westview, 1988.

Jensen, N.F. "Historical Perspectives on Plant Breeding Methodology". K.J. Frey, ed., *Historical Perspectives in Plant Science*. Ames: Iowa State University Press, 1994, pp. 179–194.

Jepma, C.J. *Tropical Deforestation: A Socio-Economic Approach*. London: Earthscan, 1995.

Jha, D.N. *The Myth of the Holy Cow*. London: Verso, 2002 [first published New Delhi: CB, 2001].

Johnson, D.H. and D.M. Anderson, eds. *The Ecology of Survival: Case Studies from Northeast African History*. London: Lester Crook, 1988.

Jonsson, U., A.-M. Köll and R. Pettersson. "What Is Wrong with a Peasant-based Development Strategy? Use and Misuse of Historical Experiences". M. Mörner and T. Svensson, eds. *The Transformation of Rural Society in the Third World*. London: Routledge, 1991, pp. 64–97.

Journal of Southern African Studies 15 (1989), Special Issue on Conservation in Southern Africa.

Kajembe, G.C. *Indigenous Management Systems as a Basis for Community Forestry in Tanzania: A Case Study of the Dodoma Urban and Lushoto Districts*. Wageningen: Wageningen Agricultural University Tropical Resource Management Papers, 1994.

Kanya-Forstner, A.S. "The French Marines and the Conquest of Western Sudan, 1880–1899". J.A. de Moor and H.L. Wesseling, eds., *Imperialism and War: Essays on Colonial Wars*. Leiden: Leiden University Press, 1989.

Kea, R. *Settlements, Trade, and Polities in the Seventeenth-Century Gold Coast*. Baltimore: Johns Hopkins University Press, 1982.

Kerkhof, P. *Agroforestry in Africa: A Survey of Project Experience*. London: The Panos Institute, 1990.

Kessy, J.F. *Conservation and Utilization of Natural Resources in the East Usambara Forest Reserve: Conventional Views and Local Perspectives*. Wageningen: Wageningen Agricultural University, 1998.

King, K.F.S. "The History of Agroforestry". Nair, *Agroforestry Systems in the Tropics*, pp. 3–11.

Kiple, K. *The Caribbean Slave: A Biological History*. Cambridge: Cambridge University Press, 2002.

Kjekhus, H. *Ecology Control and Economic Development in East African History: The Case of Tanganyika, 1850–1950*. London: Heinemann, 1977.

Klieman, K.A. *"The Pygmies Were Our Compass": Bantu and Batwa in the History of West Central Africa, Early Times to c. 1900 CE*. Portsmouth: Heinemann, 2003.

Knapen, H. *Forests of Fortune? The Environmental History of Southeast Borneo, 1600–1880*. Leiden: KITLV Press, 2001.

Knight, J. "From Timber to Tourism: Recommoditizing the Japanese Forest". Doornbos, South and White, *Forests*, pp. 333–350.

Konrad, H.W. "Tropical Forest Policy and Practice during the Mexican Porfirato, 1876–1910". Steen and Tucker, *Changing Tropical Forests*. pp. 123–143.

Koponen, J. "Population: A Dependent Variable". Maddox, Giblin and Kimambo, *Custodians of the Land*, pp. 19–42.

Kozlowski, T.T., P.J. Kramer and S.G. Pallardy. *The Physiological Ecology of Woody Plants*. San Diego: Academic Press, 1991.

Krech, S., III. *The Ecological Indian: Myth and History*. New York: Norton, 1999.

Kreike, E. "Early Asante and the Struggle for Economic and Political Control on the Gold Coast, 1690–1730". MA thesis, University of Amsterdam, 1986.

[Kreike, E.] "Historical Dynamics of Land Tenure in Ovamboland". NEPRU Working Paper no. 2 for the National Land Reform Conference, Windhoek, Namibia, June 1991.

Kreike, E. "An Inventory of Trials with Exotic Tree Species in Northern Namibia, with Special Reference to Provenance Trials with Eucalyptus spp". Windhoek: Directorate of Forestry Internal Report, 1992.

Kreike, E. "The Ovambo Agro-Silvipastoral System: Traditional Land Use and Indigenous Natural Resource Management in Northcentral Namibia". Windhoek: Ministry of Environment and Tourism, Forestry Publication No. 4, 1995.

Kreike, E. "Recreating Eden: Agro-Ecological Change, Food Security, and Environmental Diversity in Southern Angola and Northern Namibia, 1890–1960". PhD diss., Yale University, 1996.

Kreike, E. "Hidden Fruits: A Social Ecology of Fruit Trees in Namibia and Angola, 1880s–1990s". W. Beinart and J. McGregor, eds., *Social History and African Environments*. Oxford: James Currey, 2003, pp. 27–42.

Kreike, E. *Re-creating Eden: Land Use, Environment, and Society in Southern Angola and Northern Namibia*. Portsmouth, N.H.: Heinemann, 2004.

Kreike, E. "Architects of Nature: Environmental Infrastructure and the Nature-Culture Dichotomy". DrSc thesis, Environmental Sciences, Wageningen Agricultural University, 2006.

Kreike, E. "The Nature-Culture Trap: A Critique of Late Twentieth Century Global Paradigms of Environmental Change in Africa and Beyond". *Global Environment: A Journal of History and Natural and Social Sciences* 1 (2008): 114–144.

Kreike, E. "De-Globalization and Deforestation in Colonial Africa: Closed Markets, the Cattle Complex, and Environmental Change in North-Central Namibia, 1890–1990". *Journal of Southern African Studies*, forthcoming.

Lau, B. *Southern and Central Namibia in Jonker Afrikaner's Time*. Windhoek: Namibia Archives, 1987.

Lau, B., ed. *Carl Hugo Hahn Tagebücher 1837–1860; Diaries: A Missionary in Nama- and Damaraland*. Windhoek: Archives Services Division, 1985.

Law, Robin. *The Horse in West African History: The Role of the Horse in the Societies of Pre-colonial West Africa*. Oxford: Oxford University Press, 1980.

Lawton, R.M. "Browse in Miombo Woodland". Le Houérou, *Browse in Africa*, pp. 25–31.

Laycock, W.A., J.A. Young and D.N. Uechert. "Ecological Status of Poisonous Plants on Rangelands". James, Ralphs and Nielsen, *The Ecology and Economic Impact of Poisonous Plants on Livestock Production*, pp. 27–42.

Le Houérou, H.N. "The Role of Browse in the Management of Natural Grazing Lands". Le Houérou, *Browse in Africa*, pp. 329–338.

Le Houérou, H.N. *The Grazing Land Ecosystems of the African Sahel*. Berlin: Springer-Verlag, 1989.

Le Houérou, H.N., ed. *Browse in Africa: The Current State of Knowledge*. Addis Ababa: ICLA, 1980.

Leach, G. and R. Mearns. *Beyond the Fuelwood Crisis: People, Land and Trees in Africa.* London: Earthscan, 1988.

Leach, M. and R. Mearns, eds. *The Lie of the Land: Challenging Received Wisdom on the African Environment.* Oxford: IAI and James Currey; Portsmouth, N.H.: Heinemann, 1996.

Leakey, R. and R. Lewin. *The Sixth Extinction: Biodiversity and Its Survival.* London: Weidenfeld and Nicolson, 1996.

Lee, R. "What Hunters Do for a Living; or How to Make Out on Scarce Resources". R. Lee and I. DeVore, eds., *Man the Hunter.* Chicago: Aldine, 1968, pp. 30–43.

Lee, R.B. *The !Kung San: Men, Women, and Work in a Foraging Society.* Cambridge: Cambridge University Press, 1979.

Lehman, M.P. "Deforestation and Changing Land Use Patterns in Costa Rica". Steen and Tucker, *Changing Tropical Forests*, pp. 58–76.

Leuschner, W.A. and K. Khaleque. "Homestead Agroforestry in Bangladesh". Nair, *Agroforestry Systems in the Tropics*, pp. 197–209.

Leyden, B.W., M. Brenner, T. Whitmore, J.H. Curtis, D.R. Piperno and B.H. Dahlin. "A Record of Long- and Short-Term Variation from Northwest Yucatán: Cenote San José Culchacá". Fedick, *The Managed Mosaic*, pp. 30–49.

Li, Mingguang et al., "Seedling Demography in Undisturbed Tropical Wet Forest in Costa Rica". M.D. Swaine, ed., *The Ecology of Tropical Forest Tree Seedlings.* Paris: UNESCO, 1996, pp. 285–314.

Lima, D.M. de. *A Campanha dos Cuamatos Contado por um Soldado Expedicionario.* Lisbon: Livraria Ferreira, 1908.

Little, P.D. "Rethinking Interdisciplinary Paradigms and the Political Ecology of Pastoralism in East Africa". Bassett and Crummey, *African Savannas*, pp. 161–177.

Longman, K.A. and J. Jeník. *Tropical Forest and its Environment.* 2d ed. Burnt Mill, UK: Longman, 1987 [first published 1974].

Lonsdale, J. "East Africa". J.M. Brown and Wm. Roger Louis, eds., *The Oxford History of the British Empire*, vol. 4: *The Twentieth Century.* Oxford: Oxford University Press, 1999, 530–544.

Lyons, M. *The Colonial Disease: A Social History of Sleeping Sickness in Northern Zaire, 1900–1940.* Cambridge: Cambridge University Press, 1992.

Maack, P.A. "'We Don't Want Terraces!' Protest and Identity under the Uluguru Land Usage Scheme". Maddox, Giblin and Kimambo, *Custodians of the Land*, pp. 152–170.

Mackenzie, A.F.D. "Contested Ground, Colonial Narratives and the Kenyan Environment, 1920–1945". *Journal of Southern African Studies* 26(4) (2000): 697–718.

MacKenzie, J.M. *The Empire of Nature: Hunting, Conservation and British Imperialism.* Manchester: Manchester University Press, 1988.

MacKenzie, J.M., ed. *Imperialism and the Natural World.* Manchester: Manchester University Press, 1990.

MacLeod, M.J. "Exploitation of Natural Resources in Colonial Central America: Indian and Spanish Approaches". Steen and Tucker, *Changing Tropical Forests*, pp. 31–39.

Maddox, G. "Environment and Population Growth in Ugogo, Central Tanzania". Maddox, Giblin and Kimambo, *Custodians of the Land*, pp. 43–65.

Maddox, G., J. Giblin and I.N. Kimambo, eds. *Custodians of the Land: Ecology and Culture in the History of Tanzania*. London: James Currey, 1996.

Malthus, T.R. *An Essay on the Principle of Population*. Edited by D. Winch. Cambridge: Cambridge University Press, 1992.

Mamdani, M. *Citizen and Subject: Contemporary Africa and the Legacy of Late Colonialism*. Princeton, N.J.: Princeton University Press, 1996.

Mandala, E.C. *Work and Control in a Peasant Economy: A History of the Lower Tchiri Valley in Malawi, 1859–1960*. Madison: University of Wisconsin Press, 1990.

Maret, P. de and G. Thiry. "How Old Is the Iron Technology in Africa?" Schmidt, *The Culture and Technology of African Iron Production*, pp. 29–39.

Marks, R. *Tigers, Rice, and Salt: Environment and Economy in Late Imperial South China*. Cambridge: Cambridge University Press, 1998.

Marks, R.B. "Commercialization without Capitalism: Processes of Environmental Change in South China, 1550–1850". *Environmental History* 1 (Jan. 1996): 56–82.

Marquardsen, H. *Angola*. Berlin: Dietrich Reimer, 1920.

Matheny, R.T. "Northern Maya Lowland Water Control Systems". Harrison and Turner, *Pre-Hispanic Maya Agriculture*, pp. 185–210.

Mazzucato, V. and D. Niemeijer. *Rethinking Soil and Water Conservation in a Changing Society: A Case Study in Eastern Burkina Faso*. Wageningen: Wageningen University, 2000.

McAllister, P.A. "Resistance to 'Betterment' in the Transkei: A Case Study from Willowvale District". *Journal of Southern African Studies* 15(2) (1989): 346–368.

McCann, J. *People of the Plow: An Agricultural History of Ethiopia, 1800–1990*. Madison: University of Wisconsin Press, 1995.

McCann, J.C. "Climate and Causation in African History". *International Journal of African Historical Studies* 32 (2–3) (1999): 261–280.

McCann, J.C. *Green Land, Brown Land, Black Land: An Environmental History of Africa, 1800–1990*. Portsmouth, N.H.: Heinemann; Oxford: James Currey, 1999.

McKell, C.M. "Multiple Use of Fodder Trees and Shrubs—A World Wide Perspective". Le Houérou, *Browse in Africa*, pp. 141–149.

McNeill, W.H. *Plagues and Peoples*. New York: Doubleday, 1998.

McShaw, T.O. and E. McShane-Caluzi. "Resource Use in Gabon: Sustainability or Biotic Impoverishment?" Reed and Barnes, *Culture, Ecology, and Politics in Gabon's Rainforest*, pp. 7–36.

McShea, W.J. and W.M. Healy. "Oaks and Acorns as a Foundation for Ecosystem Management". McShea and Healy, *Oak Forest Ecosystems*, pp. 1–9.

McShea, W.J. and W.M. Healy, eds. *Oak Forest Ecosystems: Ecology and Management for Wildlife*. Baltimore: Johns Hopkins University Press, 2002.

Meggers, B.J. "Natural Versus Anthropogenic Sources of Amazonian Biodiversity: The Continuing Quest for El Dorado". Bradshaw and Marquet, *How Landscapes Change*, pp. 89–107.

Melville, E.K. *A Plague of Sheep: Environmental Consequences of the Conquest of Mexico*. Cambridge: Cambridge University Press, 1997.

Mendelsohn, J., S. el Obeid and C. Roberts. *A Profile of North-Central Namibia*. Windhoek: Gamsberg Macmillan, n.d.

Merchant, C. *Ecological Revolutions: Nature, Gender, and Science in New England*. Chapel Hill: University of North Carolina Press, 1989.

Merchant, C. *Reinventing Eden: The Fate of Nature in Western Culture*. New York: Routledge, 2003.

Miehr, S. "Acacia albida and Other Multipurpose Trees on the Fur Farmlands in the Jebel Marra Highlands, Western Darfur, Sudan". Nair, *Agroforestry Systems in the Tropics*, pp. 353–384.

Misana, S., C. Mung'ong'o and B. Mukamuri. "Miombo Woodlands in the Wider Context: Macro-Economic and Inter-Sectoral Influences" Campbell, *The Miombo in Transition*, pp. 73–99.

Möller, P. *Journey in Africa through Angola, Ovampoland and Damaraland*. Cape Town: C. Struik, 1974 [translated from the original Swedish edition of 1899].

Moore, H.L. and M. Vaughan. *Cutting Down Trees: Gender, Nutrition, and Agricultural Change in the Northern Province of Zambia, 1890–1990*. Portsmouth, N.H.: Heinemann, 1994.

Munro, W.A. "Ecological 'Crisis' and Resource Management Policy in Zimbabwe's Communal Lands". Bassett and Crummey, *African Savannas*, pp. 178–204.

Murray, C. *Families Divided: The Impact of Migrant Labour in Lesotho*. Johannesburg: Ravan Press, 1981.

Murray, J.A. *Wild Africa: Three Centuries of Nature Writing from Africa*. New York: Oxford University Press, 1993.

Mutwira, R. "A Question of Condoning Game Slaughter: Southern Rhodesian Wildlife Policy, 1890–1953". *Journal of Southern African Studies* 15(2) (1989): 250–262.

Myers, N. *Deforestation Rates in Tropical Forests and Their Climatic Implications*. London: Friends of the Earth Trust, 1991 [reprint; first published 1989].

Nair, P.K.R. "Agroforestry Defined". Nair, *Agroforestry Systems in the Tropics*, pp. 14–18.

Nair, P.K.R., ed. *Agroforestry Systems in the Tropics*. Dordrecht: Kluwer, 1989.

Nash, R. *Wilderness in the American Mind*. New Haven, Conn.: Yale University Press, 1982 [first published, 1967].

Nash, R.F. *American Environmentalism: Readings in Conservation History*. New York: McGraw Hill, 1990.

NEPRU [Namibian Economic Policy Research Unit]. "Land related Issues in the Communal Areas, 1: Owambo". Windhoek: Paper for the National Land Conference, 1991.

Neumann, R.P. *Imposing Wilderness: Struggles over Livelihood and Nature Preservation in Africa*. Berkeley: University of California Press, 2000 [first published 1998].

Notkola, V. and H. Siiskonen. *Fertility, Mortality and Migration in SubSaharan Africa: The Case of Ovamboland in North Namibia, 1925–90*. Houndsmills, UK: Macmillan, 2000.

Núñez, L. and M. Grosjean. "Biodiversity and Human Impact During the Last 11,000 Years in North-Central Chile". Bradshaw and Marquet, *How Landscapes Change*, pp. 7–17.

Nuttall, S. and C. Coetzee, eds. *Negotiating the Past: The Making of Memory in South Africa*. Oxford: Oxford University Press, 1999 [reprint; first published 1998].

Okafor, J.C. and E.C.M. Fernandes. "The Compound Farms of South-Eastern Nigeria: A Predominant Agroforestry Homegarden System with Crops and Small Livestock". Nair, *Agroforestry Systems in the Tropics*, pp. 411–426.

Palmer, R. and N. Parsons, eds. *The Roots of Rural Poverty in Central and Southern Africa*. Berkeley: University of California Press, 1977.

Pankhurst, R. and D.H. Johnson. "The Great Drought and Famine of 1888–92 in Northeast Africa". Johnson and Anderson, *The Ecology of Survival*, pp. 47–70.

Patterson, K.D. *Health in Colonial Ghana: Disease, Medicine, and Socio-Economic Change, 1900–1955*. Waltham, Mass.: Crossroads, 1981.

Peires, J.P. *The Dead Will Arise: Nongqawuse and the Great Xhosa Cattle killing Movement of 1856–7*. Johannesburg: Ravan Press, 1989.

Peluso, N.L. *Rich Forests, Poor People: Resource Control and Resistance in Java*. Berkeley: University of California Press, 1994 [first published 1992].

Perera, V. and R.D. Bruce. *The Last Lords of Palenque: The Lacandon Indians of the Mexican Rainforest*. Berkeley: University of California Press, 1985 [first published 1982].

Person, Y. *Une révolution Dyula*. 3 vols. Dakar, Senegal: IFAN, 1968–1975.

"A Peste Bovina em Angola". *Portugal em África* 5(52) (March 1898): 128–136.

Phimister, I. "Discourse and the Discipline of Historical Context: Conservationism and Ideas about Development in Southern Rhodesia". *Journal of Southern African Studies* 12(2) (1986): 263–275.

Pierce, S.M. "Environmental History of La Selva Biological Station: How Colonization and Deforestation of Sarapiquí Canton, Costa Rica, have altered the Ecological Context of the Station". Steen and Tucker, *Changing Tropical Forests*, pp. 40–57.

Pimentel, D.L., L. Westra and R.F. Noss, eds. *Ecological Integrity: Integrating Environment, Conservation, and Health*. Washington, D.C.: Island Press, 2000.

Pingali, P., Y. Bigot and H.P. Binswanger. *Agricultural Mechanization and the Evolution of Farming Systems in Sub-Saharan Africa*. Baltimore: John Hopkins University Press, 1987.

Piot, J. "Management and Utilization Methods for Ligneous Forages: Natural Stands and Artificial Plantations". Le Houérou, *Browse in Africa*, pp. 339–349.

Pitman, W.D. "Environmental Constraints to Tropical Forage Plant Adaptation and Productivity". Sotomayor-Ríos and Pitman, *Tropical Forest Plants*, pp. 17–23.

Pollard, J. "The Importance of Deterrence: Responses of Grazing Animals to Plant Variation". R.S. Fritz and E.L. Simms, eds., *Plant Resistance to Herbivores and Pathogens: Ecology, Evolution, and Genetics*. Chicago: University of Chicago Press, 1992, pp. 216–239.

Puleston, D.E. "Terracing, Raised Fields, and Tree Cropping in the Maya Lowlands: A New Perspective on the Geography of Power". Harrison and Turner, *Pre-Hispanic Maya Agriculture*, pp. 225–245.

Pyburn, K.A. "The Political Economy of Ancient Maya Land Use: The Road to Ruin". Fedick, *The Managed Mosaic*, pp. 236–247.

Pyne, S. *Vestal Fire: An Environmental History Told through Fire of Europe and Europe's Encounter with the World*. Seattle: University of Washington Press, 1997.

Pyne, S. *World Fire: The Culture of Fire on Earth*. Seattle: University of Washington Press, 1997.

Quadros Flores, A. de. *Recordações do Sul de Angola, 1914–1929*. Guimarães, Depositoria: Livraria L. Oliveira, [1952].

Quisumbing, A.R. and K. Otsuka, with S. Suyanto, J.B. Aidoo and E. Payongayong. *Land, Trees and Women: Evolution of Land Tenure Institutions in Western Ghana and Sumatra*. Washington, D.C.: International Food Policy Research Institute, 2001.

Rackham, O. *Trees and Woodland in the British Landscape*. Rev. ed. London: J.M. Dent, 1993 [first published 1976].

Ralphs, M.H. and L.A. Sharp. "Management to Reduce Livestock Loss from Poisonous Plants". James, Ralphs and Nielsen, *The Ecology and Economic Impact of Poisonous Plants on Livestock Production*, pp. 391–405.

Ranger, T. "Whose Heritage? The Case of the Matobo National Park". *Journal of Southern African Studies* 15(2) (1989): 217–249.

Rau, B. *From Feast to Famine: Official Cures and Grassroots Remedies to Africa's Food Crisis*. London: Zed, 1993 [reprint; first published 1991].

Reed, M.C. and J.F. Barnes, eds. *Culture, Ecology, and Politics in Gabon's Rainforest*. Lewiston, N.Y.: Edwin Mellen, 2003.

Republic of Namibia. *1991 Population and Housing Census*. Windhoek: Central Statistical Office, 1993.

Richards, P. *Indigenous Agricultural Revolution: Ecology and Food Production in West Africa*. London: Hutchinson; Boulder, Colo.: Westview Press, 1985.

Rietbergen, S., ed. *The Earthscan Reader on Tropical Forestry*. London: Earthscan, 1993.

Rittenhouse, L.R. "Toxicity Problems Associated with the Grazing of Oak in Intermountain and Southwestern USA". James, Ralphs and Nielsen, *The Ecology and Economic Impact of Poisonous Plants on Livestock Production*, pp. 197–206.

Rocheleau, D.E., P.E. Steinberg and P.A. Benjamin. "A Hundred Years of Crisis? Environment and Development Narratives in Ukwambani, Kenya". Boston: Boston University African Studies Center Working Papers, 1994.

Rodin, R.J. *The Ethnobotany of the Kwanyama Ovambos*. St. Louis: Missouri Botanical Garden, 1985.

Rolls, E. *They All Ran Wild: The Animals and Plants That Plague Australia*. London: Angus & Robertson, 1984.

Ross, R. *Adam Kok's Griquas: A Study in the Development of Stratification in South Africa*. Cambridge: Cambridge University Press, 1976.

Sahlins, M. *Stone Age Economics*. Chicago: Aldine-Atherton, 1972.

Salokoski, M. "Symbolic Power of Kings in Pre-Colonial Ovambo Societies". Licenciate Thesis in Sociology / Social Anthropology, University of Helsinki, 1992.

Sanford, S. *Management of Pastoral Development in the Third World*. Chichester, UK: Wiley, 1983.

Sauer, C. *Seeds, Spades, Hearths, and Herds: The Domestication of Animals and Foodstuffs*. Cambridge, Mass.: MIT Press, 1972.

Scarborough, V. "Reservoirs and Watersheds in the Central Maya Lowlands". Fedick, *The Managed Mosaic*, pp. 304–314.

Schachtzabel, A.P.G. *Angola: Forschungen und Erlebnisse in Südwestafrika*. Berlin: Die Buchgemeinde, 1926.

Schama, S. *Landscape and Memory*. New York: Alfred Knopf, 1995.

Scherr, S.J. and S. Franzel. "Promoting Agroforestry Technologies: Policy Lessons from On-Farm Research". Franzel and Scherr, *Trees on the Farm*, pp. 145–164.

Schmidt, P.R., ed. *The Culture and Technology of African Iron Production*. Gainesville: University Press of Florida, 1996.

Schmidt, P.R. and D.H. Avery. "Complex Iron Smelting and Pre-Historic Culture in Tanzania". Schmidt, *The Culture and Technology of African Iron Production*, pp. 172–185.

Schroeder, R.A. "Shady Practice: Gender and the Political Ecology of Resource Stabilization in the Gambian Garden / Orchards". *Economic Geography* 69(4) (1993): 349–365.

Schroth, G. and F.L. Sinclair, eds. *Trees, Crops and Soil Fertility: Concepts and Research Methods*. Wallingford, UK: CABI, 2003.

Scoones, I. "Range Management Science and Policy: Politics, Polemics and Pasture in Southern Africa". Leach and Mearns, *The Lie of the Land*, pp. 34–53.

Scott, J. *The Moral Economy of the Peasant: Rebellion and Subsistence in Southeast Asia*. New Haven, Conn.: Yale University Press, 1976.

Seely, M. and A. Marsh, eds. *Oshanas: Sustaining People, Environment, and Development in Central Ovambo*. [Windhoek], 1992.

Seward, B.R.T. "Threatened Dry-Forest Ecosystems". G.D. Piearce and D.J. Gumbo, eds., P. Shaw, comp. *The Ecology and Management of Indigenous Forests in Southern Africa: Proceedings of an International Symposium, Victoria Falls, Zimbabwe, 27–29 July 1992*. Harare: The Forestry Commission in collaboration with SAREC, 1993, pp. 273–278.

Shelton, H.M. and J.L. Brewbaker. "*Leucaena leucocephala*: The Most Widely Used Forage Tree Legume". R.C. Gutteridge and H.M. Shelton, eds., *Forage Tree Legumes in Tropical Agriculture*. Wallingford, UK: CABI, 1994, pp. 97–108.

Sheperd, G., E. Shanks and M. Hobley. "Management of Tropical and Subtropical Dry Forests". Rietbergen, *The Earthscan Reader on Tropical Forestry*, pp. 107–136.

Showers, K.B. "Soil Erosion in the Kingdom of Lesotho: Origins and Colonial Response, 1830s–1950s". *Journal of Southern African Studies* 15(2) (1989): 263–286.

Showers, K.B. *Imperial Gullies: Soil Erosion and Conservation in Lesotho*. Athens: Ohio University Press, 2005.

Siebert, S.F. "Beyond Malthus and Perverse Incentives: Economic Globalization, Forest Conversion and Habitat Fragmentation". Bradshaw and Marquet, *How Landscapes Change*, pp. 19–32.

Simon, S.F. "Sustainable Development: Theoretical Construct or Attainable Goal?" *Environmental Conservation* 16(1) (1989): 41–48.

Soini, S. "Agriculture in Northern Namibia, Owambo and Kawango 1965–1970". *Journal of the Scientific Agricultural Society of Finland* 53(3) (1981):168–209.

Sotomayor-Ríos, A. and W.D. Pitman, eds. *Tropical Forest Plants: Development and Use*. Boca Raton, Fla.: CRC Press, 2001.

Spear, T. *Mountain Farmers: Moral Economies of Land and Agricultural Development in Arusha and Mweru*. Dar es Salaam: Mkuki na Nyota, 1997.

Spear, T. and R. Waller, eds. *Being Maasai: Ethnicity and Identity in East Africa*. London: J. Currey, 1993.

Sponsel, L.E. "The Environmental History of Amazonia: Natural and Human Disturbance, and the Ecological Transition". Steen and Tucker, *Changing Tropical Forests*, pp. 233–251.

Sponsel, L.E., T.N. Headland and R. Baily. "Anthropological Perspectives on the Causes, Consequences, and Solutions of Deforestation". Sponsel, Headland and Baily, *Tropical Deforestation*, pp. 1–52.

Sponsel, L.E., T.N. Headland and R. Baily, eds. *Tropical Deforestation: The Human Dimension*. New York: Columbia University Press, 1996.

Steen, H.K. and R.P. Tucker, eds. *Changing Tropical Forests: Historical Perspectives on Today's Challenges in Central and South America*. Durham, N.C.: Forest History Society, 1992.

Steenkamp, W. *Borderstrike: South Africa Hits SWAPO Bases in Angola*. Durban: Butterworths, 1983.

Stein, S. *Vassouras: A Brazilian Coffee County, 1850–1900. The Roles of Planter and Slave in a Plantation Society*. Princeton, N.J.: Princeton University Press, 1985.

Steinhart, E.I. "Hunters, Pachers and Gamekeepers: Towards a Social History of Hunting in Colonial Kenya". *Journal of African History* 30(2) (1989): 247–264.

Stevinson-Hamilton, J. *South African Eden: The Kruger National Park, 1902–1946*. Cape Town: Struik, 1993 [first published 1937].

Stilgoe, J.R. *Common Landscape of America, 1580–1845*. New Haven, Conn.: Yale University Press, 1982.

Stuart, G.S. and G.E. Stuart. *Lost Kingdoms of the Maya*. Washington D.C.: The Society, 1993.

Sundan, N. "Unpacking the 'Joint' in Joint Forest Management". Doornbos, South and White, *Forests*, pp. 249–273.

Suret-Canale, J. *Afrique Noire: L'ère coloniale, 1900–1945*. Paris: Editions Sociales, 1964.

Swart, S. " 'Horses! Give Me More Horses!': White Settler Identity, Horses, and the Making of Early Modern South Africa". K. Raber and T.J. Tucker, eds., *The Culture of the Horse: Status, Discipline, and Identity in the Early Modern World*. New York: Palgrave, 2005, pp. 311–328.

Swift, D.M., M.B. Conghenour and M. Atsedu. "Arid and Semi-Arid Ecosystems". T.R. McClanahan and T.P. Young, eds., *East African Ecosystems and Their Conservation*. New York: Oxford University Press, 1996, pp. 243–272.

Swift, J. "Desertification: Narratives, Winners and Losers". Leach and Mearns, *The Lie of the Land*, pp. 73–90.

Taylor, S. *Shaka's Children: A History of the Zulu People*. London: HarperCollins, 1994.

Thomas, K. *Man and the Natural World: Changing Attitudes in England, 1500–1800*. New York: Oxford University Press, 1996 [first published 1983].

Thompson, E. *The Soundscape of Modernity: Architectural Acoustics and the Culture of Listening in America, 1900–1933*. Cambridge, Mass.: MIT Press, 2004.

Tiffen, M., M. Mortimore and F. Gichuki. *More People, Less Erosion: Environmental Recovery in Kenya*. Chichester, UK: Wiley, 1994.

Tomlinson, P.B. and M.H. Zimmermann, eds. *Tropical Trees as Living Systems*. Cambridge: Cambridge University Press, 1978.

Toure, O. "The Pastoral Environment of Northern Senegal". *Review of African Political Economy* 15(42) (1988): 32–39.

Trefon, T. "Libreville's Evolving Forest Dependencies". Reed and Barnes, *Culture, Ecology, and Politics in Gabon's Rainforest*, pp. 37–62.

Tucker, R.P. "The Depletion of India's Forests under British Imperialism: Planters, Foresters and Peasants in Assam and Kerala". Worster, *The Ends of the Earth*, pp. 118–140.

Turvey, B.H.C., comp., and W. Zimmermann and G.B. Taapopi, eds. *Kwanyama-English Dictionary*. Johannesburg: Witwatersrand University Press, 1977.

Vail, L. "Ecology and History: The Example of Eastern Zambia". *Journal of Southern African Studies* 3 (1977): 129–155.

Vail, L., ed. *The Creation of Tribalism in Southern Africa*. London: James Currey, 1989.

Van Beek, W.E.A. and P.M. Banga. "The Dogon and their Trees". D. Parkin and E. Croll, eds. *Bush Base: Forest Farm: Culture, Environment and Development*. London: Routledge, 1992, pp. 57–75.

Van den Bergh, J. "Diverging Perceptions on the Forest: Bulu Forest Tenure and the 1994 Cameroon Forest Law". Wiersum, *Tropical Forest Resource Dynamics*, pp. 99–114.

Van der Haar, G. "Peasant Control and the Greening of the Tojolabal Highlands, Mexico". Wiersum, *Tropical Forest Resource*, pp. 99–114.

Van Helden, F. "Resource Dynamics, Livelihood and Social Change on the Forest Fringe: A View from the Highlands of Papua New Guinea". Wiersum, *Tropical Forest Resource Dynamics*, pp. 71–98.

Van Onselen, C. "Reactions to Rinderpest in Southern Africa, 1896–97". *Journal of African History* 13(3) (1972): 473–488.

Van Sittert, L. "'The Seed Blows About in Every Breeze': Noxious Weed Eradication in the Cape Colony, 1860–1909". *Journal of Southern African Studies* 26(4) (2000): 655–674.

Vandermeer, J. "The Human Niche and Rain Forest Preservation in Southern Central America". Sponsel, Headland and Baily, *Tropical Deforestation*, pp. 216–229.

Vansina, J. *Paths in the Rainforest: Toward a History of Political Tradition in Equatorial Africa*. Madison: Wisconsin University Press, 1990.

Varughese, G. "Population and Forest Dynamics in the Hills of Nepal: Institutional Remedies by Rural Communities". Gibson, McKean, and Ostrom, *People and Forests*, pp. 193–226.

Venning, J.H. "Notes on Southern Rhodesian Ruins in Victoria District". *Journal of the Royal African Society* 7(26) (Jan. 1908): 150–158.

Veth, P.J. *Daniel Veth's Reizen in Angola*. Haarlem: Tjeenk Willink, 1887.

Villiers, B. de. *Land Claims and National Parks: The Makuleke Experience*. Pretoria: Human Sciences Research Council, 1999.

Vlcek, D.T., S. Garza de Gonzalez and E.B. Kurjack. "Contemporary Farming and Ancient Maya Settlement: Some Disconcerting Evidence". Harrison and Turner, *Pre-Hispanic Maya Agriculture*, pp. 211–223.

Walker, B.H. "A Review of Browse and Its Role in Livestock Production in Southern Africa". Le Houérou, *Browse in Africa*, pp. 7–24.

Walker, B.L. *The Conquest of Ainu Lands: Ecology and Culture in Japanese Expansion, 1590–1800*. Berkeley: University of California Press, 2001.

Waller, R. "Tsetse Fly in Western Narok, Kenya". *Journal of African History* 31(1) (1990): 81–101.

Wambugu, C., et al. "Scaling Up the Use of Fodder Shrubs in Central Kenya". Franzel, Cooper, Denning and Eade, *Development and Agroforestry*, pp. 107–116.

Watkins, C., ed. *European Woods and Forests: Studies in Cultural History*. Wallingford, UK: CABI, 1998.

Watts, M. *Silent Violence: Food, Famine, and Peasantry in Northern Nigeria*. Berkeley: University of California Press, 1983.

Webb, J.L.A. *Desert Frontier: Ecological and Economic Change along the Western Sahel, 1600–1850*. Madison: University of Wisconsin Press, 1995.

Weber, J.C., et al. "Participatory Domestication of Agroforestry Trees: An Example from the Peruvian Amazon". Franzel, Cooper, Denning and Eade, *Development and Agroforestry*, pp. 24–34.

Wellington, J.H. "O Rio Cunene e a Planície de Etosha". *Boletim de Sociedade de Geografia* 64 (1–2) (Jan. – Feb. 1946): 459–480.

Westoby, J. *Introduction to World Forestry*. Oxford: Blackwell, 1989.

Westra, L. et al. "Ecological Integrity and the Aims of the Global Integrity Project". Pimentel, Westra and Noss, *Ecological Integrity*, pp. 19–41.

Westra, L. and P.S. Wenz, eds. *Faces of Environmental Racism: Confronting Issues of Global Justice*. Lanham, Md.: Rowman & Littlefield, 1995.

White, R. *The Middle Ground: Indians, Empires, and Republics in the Great Lakes Region, 1650–1815*. Cambridge: Cambridge University Press, 1991.

White, R. *The Organic Machine: The Remaking of the Columbia River*. New York: Hill & Wang, 2000 [first published 1995].

Wiersum, K.F. *Social Forestry: Changing Perspectives in Forest Science or Practice?* Wageningen: Wageningen Agricultural University, 1999.

Wiersum, K.F. "Use and Conservation of Biodiversity in East African Forested Landscapes". P.A. Zuidema, ed., *Tropical Forests in Multi-Functional Landscapes: Proceedings of Two Seminars Organised by the Prince Bernard Centre for International Nature Conservation, Utrecht University, in Collaboration with the Dutch Association for Tropical Foresters, held in Utrecht, 2 December 2002 and 11 April 2003*. Utrecht: Utrecht University, 2003, pp. 33–39.

Wiersum, K.F., ed. *Tropical Forest Resource Dynamics and Conservation: From Local to Global Issues*. Wageningen: Wageningen Agricultural University, 2000.

Wiersum, K.F. and G.A. Persoon. "Research on Conservation and Management of Tropical Forests: Contributions from Social Sciences in the Netherlands". Wiersum, *Tropical Forest Resource Dynamics*, pp. 1–24.

Williams, M. *Deforesting the Earth: From Prehistory to Global Crisis*. Chicago: University of Chicago Press, 2003.

Wilmsen, E.N. *Land Filled with Flies: A Political Economy of the Kalahari*. Chicago: University of Chicago Press, 1989.

Wilson, K.B. "Trees in Fields in Southern Zimbabwe". *Journal of Southern African Studies* 15(2) (1989): 369–383.

Wingard, J.D. "Interactions between Demographic Processes and Soil Resources". Fedick, *The Managed Mosaic*, pp. 207–235.

Woodwell, G.M., with contributions from O. Ullstein, R.A. Houghton, S. Nilsson, P. Kanowski, E.D. Larson, T.B. Johansson and B. Kerr. *Forests in a Full World*. New Haven, Conn.: Yale University Press, 2001.

Worster, D. *Dust Bowl: The Southern Plains in the 1930s*. Oxford: Oxford University Press, 1982 [first published 1979].

Worster, D., ed. *The Ends of the Earth: Perspectives on Modern Environmentalism*. Cambridge: Cambridge University Press, 1999 [first published 1988].

Wülfhorst, A. *Aus den Anfangstagen der Ovambo-Mission.* Barmen: Verlag des Missionshauses, 1904.

Wülfhorst, A. *Shiwesa, Ein Simeon aus den Ovambochristen.* Barmen: Verlag des Missionshauses, 1912.

Wülfhorst, A. *Von Hexen und Zaubern: Bilder aus dem Leben der heidnischen Amboleute in Süd-West-Afrika.* Wuppertal-Barmen: n.p., 1935.

Wülfhorst, A. *Moses, Eine Erstlingsfrucht aus dem Ovambo.* Wuppertal-Barmen: Missionshaus Verlag, n.d.

Young, A. *Agroforestry for Soil Management.* 2d ed. Wallingford, UK: CABI, 1997 [first published 1989].

Index

www.ingramcontent.com/pod-product-compliance
Lightning Source LLC
Chambersburg PA
CBHW031553280326
41928CB00047BA/214